DNA Technology in Forensic Science

Committee on DNA Technology in Forensic Science
Board on Biology
Commission on Life Sciences
National Research Council

NATIONAL ACADEMY PRESS
Washington, D.C. 1992

National Academy Press • 2101 Constitution Avenue., N.W. • Washington, DC 20418

NOTICE: The project that is the subject of this report was approved by the Governing Board of the National Research Council, whose members are drawn from the councils of the National Academy of Sciences, the National Academy of Engineering, and the Institute of Medicine. The members of the committee responsible for the report were chosen for their special competences and with regard for appropriate balance.

This report has been reviewed by a group other than the authors according to procedures approved by a Report Review Committee consisting of members of the National Academy of Sciences, the National Academy of Engineering, and the Institute of Medicine.

This Board on Biology study was supported by the Federal Bureau of Investigation, the National Institutes of Health National Center for Human Genome Research, the National Institute of Justice, the National Science Foundation, the Alfred P. Sloan Foundation, and the State Justice Institute.

Library of Congress Cataloging-in-Publication Data

National Research Council (U.S.). Committee on DNA Technology in Forensic Science.

DNA technology in forensic science / Committee on DNA Technology in Forensic Science, Board on Biology, Commission on Life Sciences, National Research Council.

p. cm.

Includes bibliographical references and index.

ISBN 0-309-04587-8

1. Forensic genetics—Congresses. 2. DNA fingerprinting—Congresses. I. Title.

[DNLM: 1. DNA—analysis. 2. Forensic Medicine—methods. W 786 N277d]

RA1057.5N37 1992

614'.1—dc20

DNLM/DLC

for Library of Congress 92-16341

CIP

This book is printed with soy ink on acid-free recycled stock. 

Printed in the United States of America

First Printing, July 1992
Second Printing, January 1993
Third Printing, May 1997
Fourth Printing, July 2000

COMMITTEE ON DNA TECHNOLOGY IN FORENSIC SCIENCE

VICTOR A. McKUSICK, *Chairman,* The Johns Hopkins Hospital, Baltimore, Maryland
PAUL B. FERRARA, Division of Forensic Sciences, Department of General Services, Richmond, Virginia
HAIG H. KAZAZIAN, The Johns Hopkins Hospital, Baltimore, Maryland
MARY-CLAIRE KING, University of California, Berkeley, California
ERIC S. LANDER, Whitehead Institute for Biomedical Research, Cambridge, Massachusetts
HENRY C. LEE, Connecticut State Police, Meriden, Connecticut
RICHARD O. LEMPERT, University of Michigan Law School, Ann Arbor, Michigan
RUTH MACKLIN, Albert Einstein College of Medicine, Bronx, New York
THOMAS G. MARR, Cold Spring Harbor Laboratory, Cold Spring Harbor, New York
PHILIP R. REILLY, Shriver Center for Mental Retardation, Waltham, Massachusetts
GEORGE F. SENSABAUGH, Jr., University of California, Berkeley, California
JACK B. WEINSTEIN, U.S. District Court, New York, Brooklyn, New York

Former Members

C. THOMAS CASKEY (Resigned December 21, 1991), Baylor College of Medicine, Houston, Texas
MICHAEL W. HUNKAPILLER (Resigned August 17, 1990), Applied Biosystems Inc., Foster City, California

National Research Council Staff

OSKAR R. ZABORSKY, Study Director; Director, Board on Biology
NORMAN GROSSBLATT, Editor
MARIETTA E. TOAL, Administrative Secretary
MARY KAY PORTER, Senior Secretary

COMMISSION ON LIFE SCIENCES

The National Academy of Sciences is a private, nonprofit, self-perpetuating society of distinguished scholars engaged in scientific and engineering research, dedicated to the furtherance of science and technology and to their use for the general welfare. Upon the authority of the charter granted to it by the Congress in 1863, the Academy has a mandate that requires it to advise the federal government on scientific and technical matters. Dr. Frank Press is president of the National Academy of Sciences.

The National Academy of Engineering was established in 1964, under the charter of the National Academy of Sciences, as a parallel organization of outstanding engineers. It is autonomous in its administration and in the selection of its members, sharing with the National Academy of Sciences the responsibility for advising the federal government. The National Academy of Engineering also sponsors engineering programs aimed at meeting national needs, encourages education and research, and recognizes the superior achievements of engineers. Dr. Robert M. White is president of the National Academy of Engineering.

The Institute of Medicine was established in 1970 by the National Academy of Sciences to secure the services of eminent members of appropriate professions in the examination of policy matters pertaining to the health of the public. The Institute acts under the responsibility given to the National Academy of Sciences by its congressional charter to be an advisor to the federal government and, upon its own initiative, to identify issues of medical care, research, and education. Dr. Kenneth I. Shine is president of the Institute of Medicine.

The National Research Council was organized by the National Academy of Sciences in 1916 to associate the broad community of science and technology with the Academy's purposes of furthering knowledge and of advising the federal government. Functioning in accordance with general policies determined by the Academy, the Council has become the principal operating agency of both the National Academy of Sciences and the National Academy of Engineering in providing services to the government, the public, and the scientific and engineering communities. The Council is administered jointly by both Academies and the Institute of Medicine. Dr. Frank Press and Dr. Robert M. White are chairman and vice chairman, respectively, of the National Research Council.

Acknowledgment and Disclaimer:

This report was supported with joint funding from the National Institute of Justice, the Federal Bureau of Investigation, and the State Justice Institute, under award #89-IJ-CX-0055 from the National Institute of Justice, Office of Justice Programs, U.S. Department of Justice. Points of view in this document are those of the authors and do not necessarily represent the official position of the U.S. Department of Justice.

Preface

In recent years, advances in the techniques for mapping and sequencing the human genome have contributed to progress in both basic biology and medicine. The applications of these techniques have not been restricted to biology and medicine, however, but have also entered forensic science. Today, methods developed in basic molecular biology laboratories can potentially be used in forensic science laboratories in a matter of months.

On the basis of its study of the mapping and sequencing of the human genome (reported in 1988), the Board on Biology and several federal agencies recognized the potential of DNA typing technology for forensic science. In particular, the Federal Bureau of Investigation, the preeminent organization in the United States for the development and application of forensic techniques, initiated an effort to develop and evaluate DNA typing in forensic applications in the mid-1980s. The first case work was performed in December 1988. Several private-sector laboratories entered the field early, and state government crime laboratories also began to offer services in DNA typing. However, as DNA typing entered the courtrooms of this country, questions appeared about its reliability and methodological standards and about the interpretation of population statistics.

By the summer of 1989, a crescendo of questions concerning DNA typing had been raised in connection with some well-publicized criminal cases, and calls for an examination of the issues by the National Research Council of the National Academy of Sciences came from the scientific and legal communities. As a response, this study was initiated in January 1990.

Because of the broad ramifications of forensic DNA typing, a number of federal agencies and one private foundation provided financial support for this study: the Federal Bureau of Investigation, the National Institutes of Health National Center for Human Genome Research, the National Institute of Justice, the National Science Foundation, the State Justice Institute, and the Alfred P. Sloan Foundation.

Many persons offered assistance to the committee and staff during this complex study. In particular, the following deserve recognition and praise for their efforts: John Hicks, Federal Bureau of Investigation; Elke Jordan and Eric Juengst, National Institutes of Health National Center for Human Genome Research; James K. Stewart, Charles B. DeWitt, Bernard V. Auchter, and Richard Laymon, National Institute of Justice; John C. Wooley, National Science Foundation; David I. Tevelin, State Justice Institute; and Michael S. Teitelbaum, Alfred P. Sloan Foundation.

I also thank the many experts who offered their advice to the committee during its briefings and open meetings. The names of those who offered testimony are given in the appendix. Additionally, I want to thank the many who wrote to me or to the National Research Council and provided valuable data and suggestions to the committee; much was gained from their input. We also acknowledge the efforts of Robert Kushen, Columbia Law School, in assisting Judge Weinstein. I also thank Della Malone, my secretary, for her help throughout. The committee thanks the reviewers of our report for many valuable comments and suggestions. Although the reviewers are anonymous to us, I personally want to thank them for their constructive comments and suggestions.

The staff of the Board on Biology deserve special praise for their efforts during the many months of intense activity. Oskar R. Zaborsky, Study Director and Director of the Board on Biology, deserves recognition for his administrative and technical contributions and for handling many complex matters. Marietta Toal, Administrative Secretary, served the committee well in logistics and the preparation of the report. The committee also thanks Mary Kay Porter for her assistance. Norman Grossblatt edited the report.

Last but not least, I thank my colleagues on the committee who served so well and unselfishly to address key issues from the perspective of their special expertise and to prepare this report in a timely fashion.

DNA typing for personal identification is a powerful tool for criminal investigation and justice. At the same time, the technical aspects of DNA typing are vulnerable to error, and the interpretation of results requires appreciation of the principles of population genetics. These considerations and concerns arising out of the felon DNA databanks and the privacy of DNA information made it imperative to develop guidelines and safeguards for the most effective and socially beneficial use of this powerful tool. We hope that our efforts will enhance understanding of the issues and serve to bring together people of good will from science, technology, law, and ethics. We hope that our report will serve well the sponsors and the general public.

Victor A. McKusick
Chairman
Committee on DNA Technology
in Forensic Science

A Statement by the Committee on DNA Technology in Forensic Science

On April 14, 1992, *The New York Times* printed an article on this report. That article seriously misrepresented the findings of the committee; in an article on April 15, the *Times* corrected the misrepresentation. To avoid any potential confusion engendered by the April 14 article, the committee provides the following clarifying statement:

We recommend that the use of DNA analysis for forensic purposes, including the resolution of both criminal and civil cases, be continued while improvements and changes suggested in this report are being made. There is no need for a general moratorium on the use of the results of DNA typing either in investigation or in the courts.

We regard the accreditation and proficiency testing of DNA typing laboratories as essential to the scientific accuracy, reliability, and acceptability of DNA typing evidence in the future. Laboratories involved in forensic DNA typing should move quickly to establish quality-assurance programs. After a sufficient time for implementation of quality-assurance programs has passed, courts should view quality control as necessary for general acceptance.

The Committee

Contents

Summary

Characterization, or "typing," of deoxyribonucleic acid (DNA) for purposes of criminal investigation can be thought of as an extension of the forensic typing of blood that has been common for more than 50 years; it is actually an extension from the typing of proteins that are coded for by DNA to the typing of DNA itself. Genetically determined variation in proteins is the basis of blood groups, tissue types, and serum protein types. Developments in molecular genetics have made it possible to study the person-to-person differences in parts of DNA that are not involved in coding for proteins, and it is primarily these differences that are used in forensic applications of DNA typing to personal identification. DNA typing can be a powerful adjunct to forensic science. The method was first used in casework in 1985 in the United Kingdom and first used in the United States by commercial laboratories in late 1986 and by the Federal Bureau of Investigation (FBI) in 1988.

DNA typing has great potential benefits for criminal and civil justice; however, because of the possibilities for its misuse or abuse, important questions have been raised about reliability, validity, and confidentiality. By the summer of 1989, the scientific, legal, and forensic communities were calling for an examination of the issues by the National Research Council of the National Academy of Sciences. As a response, the Committee on DNA Technology in Forensic Science was formed; its first meeting was held in January 1990. The committee was to address the general applicability and appropriateness of the use of DNA technology in forensic science, the need to develop standards for data collection and analysis, aspects of the technology, management of DNA typing data, and legal, societal, and ethical issues surrounding DNA typing. The techniques of DNA typing are fruits of the revolution in molecular biology that is yielding an explosion of information about human genetics. The highly personal and sensitive information that can be generated by DNA typing requires strict confidentiality and careful attention to the security of data.

DNA, the active substance of the genes, carries the coded messages of heredity in every living thing: animals, plants, bacteria, and other microorganisms. In humans, the code-carrying DNA occurs in all cells that have a nucleus, including white blood cells, sperm, cells surrounding hair roots, and cells in saliva. These would be the cells of greatest interest in forensic studies.

Human genes are carried in 23 pairs of chromosomes, long threadlike or rodlike structures that are a person's archive of heredity. Those 23 pairs, the total genetic makeup of a person, are referred to as the human "diploid genome." The chemistry of DNA embodies the universal code in which the messages of heredity are transmitted. The genetic code itself is spelled out in strings of nucleotides of four types, commonly represented by the letters A, C, G, and T (standing for the bases adenine, cytosine, guanine, and thymine), which in various combinations of three nucleotides spell out the

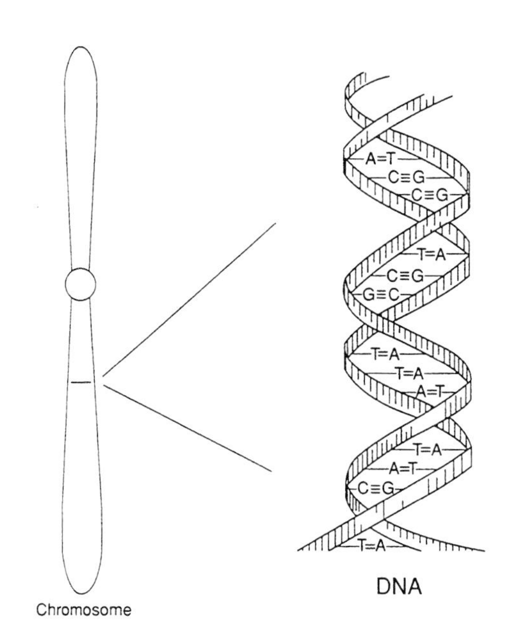

FIGURE 1 Diagram of the double-helical structure of DNA in a chromosome. The line shown in the chromosome is expanded to show the DNA structure.

codes for the amino acids that constitute the building blocks of proteins. A gene, the basic unit of heredity, is a sequence of about 1,000 to over 2 million nucleotides. The human genome, the total genetic makeup of a person, is estimated to contain 50,000-100,000 genes.

The total number of nucleotides in a set of 23 chromosomes—one from each pair, the "haploid genome"—is about 3 billion. Much of the DNA, the part that separates genes from one another, is noncoding. Variation in the genes, the coding parts, are usually reflected in variations in the proteins that they encode, which can be recognized as "normal variation" in blood type or in the presence of such diseases as cystic fibrosis and phenylketonuria; but variations in the noncoding parts of DNA have been most useful for DNA typing.

Except for identical twins, the DNA of a person is for practical purposes unique. That is because one chromosome of each pair comes from the father and one from the mother; which chromosome of a given pair of a parent's chromosomes that parent contributes to the child is independent of which chromosome of another pair that parent gives to that child. Thus, the different combinations of chromosomes that one parent can give to one child is 2^{23}, and the number of different combinations of paired chromosomes a child can receive from both parents is 2^{46}.

The substitution of even one nucleotide in the sequence of DNA is a variation that can be detected. For example, a variation in DNA consisting of the substitution of one nucleotide for another (such as the substitution of a C for a T) can often be recognized by a change in the points at which certain biological catalysts called "restriction enzymes" cut the DNA. Such an enzyme cuts DNA whenever it encounters a specific sequence of nucleotides that is peculiar to the enzyme. For example, the enzyme *Hae*I cuts DNA wherever it encounters the sequence AGGCCA. A restriction enzyme will cut a sample of DNA into fragments whose lengths depend on the location of the cutting sites recognized by the enzyme. Assemblies of fragments of different lengths are called "restriction fragment length polymorphisms" (RFLPs), and RFLPs constitute one of the most important tools for analyzing and identifying samples of DNA.

An important technique used in such analyses is the "Southern blot," developed by Edwin Southern in 1975. A sample of DNA is cut with a restriction enzyme, and the fragments are separated from one another by electrophoresis (i.e., they are separated by an electrical field). The fragments of particular interest are then identified with a labeled probe, a short segment of single-stranded DNA containing a radioactive atom, which hybridizes (fuses) to the fragments of interest because its DNA sequence is complementary to those of the fragments (A pairing with T, C pairing with G). Each electrophoretic band represents a separate fragment of DNA, and a given person will have no more than two fragments derived from a particular place in his or her DNA—one representing each of the genes that are present on the two chromosomes of a given pair. The forms of a given gene are referred to as alleles. A person who received the same allele from the mother and the father is said to be "homozygous" for that allele; a person who received different alleles from the mother and the father is said to be "heterozygous." Many RFLP systems are based on change in a single nucleotide. They are said to be "diallelic," because there are only two common alternative forms. And there are only three genotypes: two kinds of homozygous genotypes and a heterozygous genotype. Another form of RFLP is generated by the presence of variable number tandem repeats (VNTRs). VNTRs are sequences, sometimes as small as two different nucleotides (such as C and A), that are repeated in the DNA. When such a structure is subjected to cutting with restriction enzymes, fragments of varied length are obtained.

It was variation of the VNTR type to which Alex Jeffreys in the United Kingdom first applied the designation "DNA fingerprinting." He used probes that recognized not one locus, but multiple loci, and "DNA fingerprinting" has come to refer particularly to multilocus, multiallele systems. A locus is a specific site of a gene on a chromosome. In the United States, in particular, single-locus probes are preferred, because their results are easier to interpret. "DNA typing" is the preferred term, because "DNA fingerprinting" is associated with multilocus systems. Discriminating power for personal identification is achieved by using several—usually at least four—single-locus, multiallelic systems.

The entire procedure for analyzing DNA with the RFLP method is diagrammed in Figure 2.

After the bands (alleles) are visualized, those in the evidence sample and the suspect sample are compared. If the bands match in the two samples, for all three or four enzyme-probe combinations, the question is: What is the probability that such a match would have occurred between the suspect and a person drawn at random from the same population as the suspect?

Answering that question requires calculation of the frequency in the population of each of the gene variants (alleles) that have been found, and the calculation requires a databank where one can find the frequency of each allele in the population. On the basis of some assumptions, so-called Hardy-Weinberg ratios can be calculated. For a two-allele system, the ratios are indicated by the expressions p^2 and q^2 for the frequency of the two homozygotes and $2pq$ for heterozygotes, p and q being the frequencies of the two alleles and $p + q$ being equal to 1. Suppose that a person is heterozygous at a locus where the frequencies of the two alleles in the population are 0.3 and 0.7. The frequency of that heterozygous genotype in the population would be $2 \times 0.3 \times 0.7 = 0.42$. Suppose, further, that at three

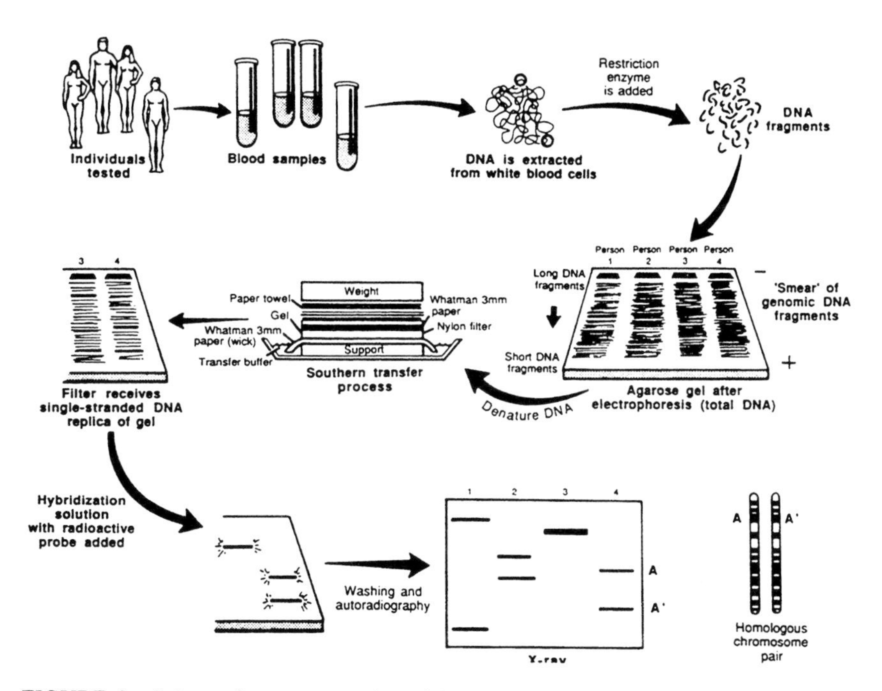

FIGURE 2 Schematic representation of Southern blotting of single-locus, multiallelic VNTR. In example shown here, DNA from four persons is tested. All have different patterns. Three are heterozygous and one homozygous, for a total of seven different alleles. From L. T. Kirby, "DNA Fingerprinting: An Introduction," Stockton Press, New York, 1990. Copyright © 1990 by Stockton Press. Reprinted with permission of W.H. Freeman and Company.

other loci the person being typed has genotypes with population frequencies of 0.01, 0.32, and 0.02. The frequency of the combined genotype in the population is $0.42 \times 0.01 \times 0.32 \times 0.02$, or 0.000027, or approximately 1 in 37,000.

That example illustrates what is called the product rule, or multiplication rule. Its use assumes that the alleles at a given locus are inherited independently of each other. It also assumes that there are no subpopulations in which a particular allele at one locus would have a preferential probability of being associated with a particular allele at a second locus.

Techniques for analyzing DNA are changing rapidly. One key technique introduced in the last few years is the polymerase chain reaction (PCR), which allows a million or more copies of a short region of DNA to be easily made. For DNA typing, one amplifies (copies) a genetically informative sequence, usually 100-2,000 nucleotides long, and detects the genotype in the amplified product. Because many copies are made, DNA typing can rely on methods of detection that do not use radioactive substances. Furthermore, the technique of PCR amplification permits the use of very small samples of tissue or body fluids—theoretically even a single nucleated cell.

The PCR process (Figure 3) is simple; indeed, it is analogous to the process by which cells replicate their DNA. It can be used in conjunction with various methods for detecting person-to-person differences in DNA.

It must be emphasized that new methods and technology for demonstrating individuality in each person's DNA are being developed. The present methods explained here will probably be superseded by others that are more efficient, error-free, automatable, and cost-effective. Care should be taken to ensure that DNA typing techniques used for forensic purposes do not become "locked in" prematurely, lest society and the criminal justice system be unable to benefit fully from advances in science and technology.

TECHNICAL CONSIDERATIONS

The forensic use of DNA typing is an outgrowth of its medical diagnostic use—analysis of disease-causing genes based on comparison of a patient's DNA with that of family members to study inheritance patterns of genes or comparison with reference standards to detect mutations. To understand the challenges involved in such technology transfer, it is instructive to compare forensic DNA typing with DNA diagnostics.

DNA diagnostics usually involves clean tissue samples from known sources. Its procedures can usually be repeated to resolve ambiguities. It involves comparison of discrete alternatives (e.g., which of two alleles did a child inherit from a parent?) and thus includes built-in consistency checks against artifacts. It requires no knowledge of the distribution of patterns in the general population.

Forensic DNA typing often involves samples that are degraded, contaminated, or from multiple unknown sources. Its procedures sometimes cannot be repeated, because there is too little sample. It often involves matching of samples from a wide range of alternatives in the population and thus lacks built-in consistency checks. Except in cases where the DNA evidence excludes a suspect, assessing the significance of a result requires statistical analysis of population frequencies.

Each method of DNA typing has its own advantages and limitations, and each is at a different state of technical development. However, the use of each method involves three steps:

1. Laboratory analysis of samples to determine their genetic-marker types at multiple sites of potential variation.

2. Comparison of the genetic-marker types of the samples to determine

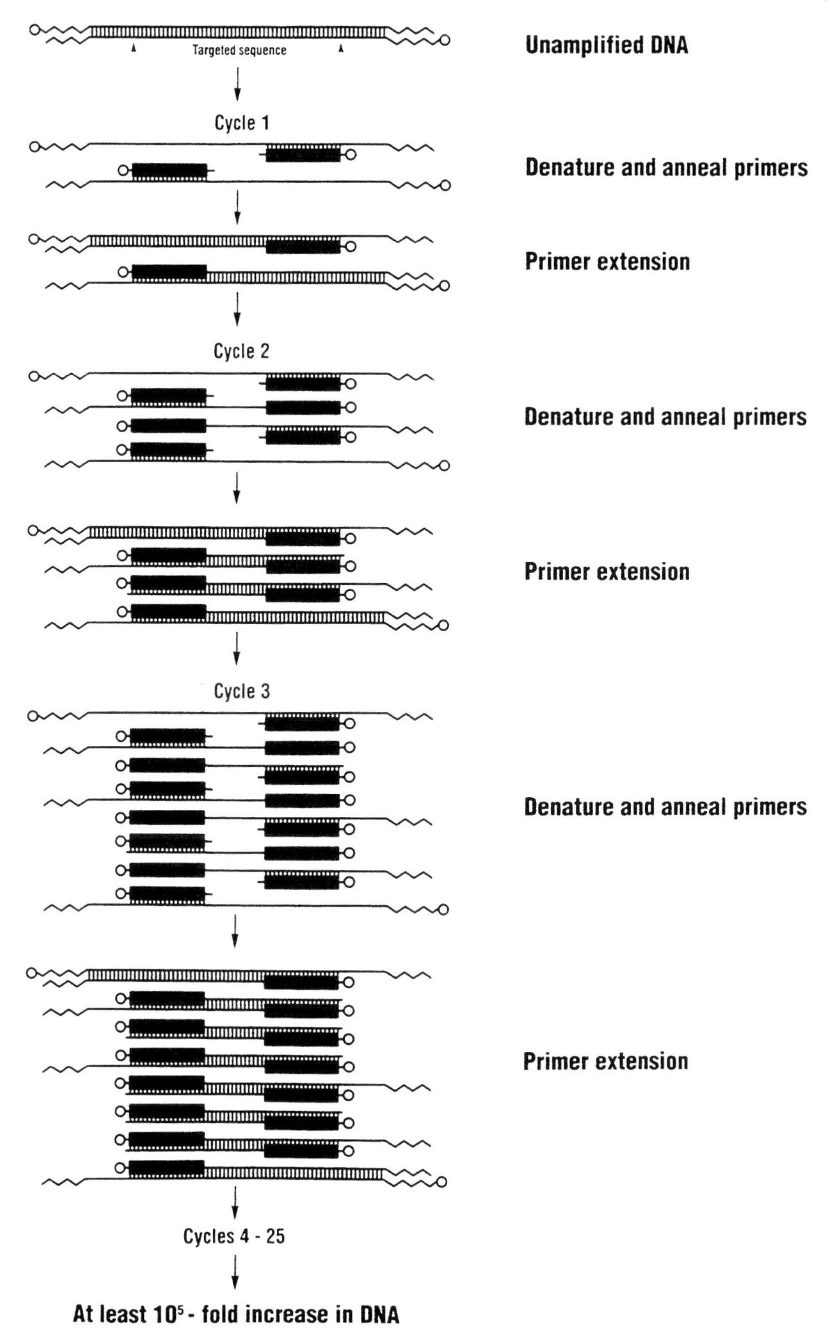

FIGURE 3 Polymerase chain reaction (PCR). Courtesy, Perkin-Elmer Cetus Instruments.

whether the types match and thus whether the samples could have come from the same source.

3. If the types match, statistical analysis of the population frequencies of the types to determine the probability that a match would have been observed by chance in a comparison of samples from different persons.

Before any particular DNA typing method is used for forensic purposes, precise and scientifically reliable procedures for performing all three steps must be established. It is meaningless to speak of the reliability of DNA typing in general—i.e., without specifying a particular method.

Despite the challenges of forensic DNA typing, it is possible to develop reliable forensic DNA typing systems, provided that adequate scientific care is taken to define and characterize the methods.

Recommendations

- Any new DNA typing method (or a substantial variation of an existing method) must be rigorously characterized in both research and forensic settings, to determine the circumstances under which it will yield reliable results.
- DNA analysis in forensic science should be governed by the highest standards of scientific rigor, including the following requirements:

  —Each DNA typing procedure must be completely described in a detailed, written laboratory protocol.

  —Each DNA typing procedure requires objective and quantitative rules for identifying the pattern of a sample.

  —Each DNA typing procedure requires a precise and objective matching rule for declaring whether two samples match.

  —Potential artifacts should be identified by empirical testing, and scientific controls should be designed to serve as internal checks to test for the occurrence of artifacts.

  —The limits of each DNA typing procedure should be understood, especially when the DNA sample is small, is a mixture of DNA from multiple sources, or is contaminated with interfering substances.

  —Empirical characterization of a DNA typing procedure must be published in appropriate scientific journals.

  —Before a new DNA typing procedure can be used, it must have not only a solid scientific foundation, but also a solid base of experience.

- The committee strongly recommends the establishment of a National Committee on Forensic DNA Typing (NCFDT) under the auspices of an appropriate government agency or agencies to provide expert advice primarily on scientific and technical issues concerning forensic DNA typing.

• Novel forms of variation in the genome that have the potential for increased power of discrimination between persons are being discovered. Furthermore, new ways to demonstrate variations in the genome are being developed. The current techniques are likely to be superseded by others that provide unambiguous individual identification and have such advantages as automatability and economy. Each new method should be evaluated by the NCFDT for use in the forensic setting, applying appropriate criteria to ensure that society derives maximal benefit from DNA typing technology.

STATISTICAL BASIS FOR INTERPRETATION

Because any two human genomes differ at about 3 million sites, no two persons (barring identical twins) have the same DNA sequence. Unique identification with DNA typing is therefore possible, in principle, provided that enough sites of variation are examined. However, the DNA typing systems used today examine only a few sites of variation and have only limited resolution for measuring the variability at each site. There is a chance that two persons have DNA patterns (i.e., genetic types) that match at the small number of sites examined. Nevertheless, even with today's technology, which uses 3-5 loci, a match between two DNA patterns can be considered strong evidence that the two samples came from the same source. Interpreting a DNA typing analysis requires a valid scientific method for estimating the probability that a random person by chance matches the forensic sample at the sites of DNA variation examined. To say that two patterns match, without providing any scientifically valid estimate (or, at least, an upper bound) of the frequency with which such matches might occur by chance, is meaningless. The committee recommends approaches for making sound estimates that are independent of the race or ethnic group of the subject.

A standard way to estimate frequency is to count occurrences in a random sample of the appropriate population and then use classical statistical formulas to place upper and lower confidence limits on the estimate. (Because forensic science should avoid placing undue weight on incriminating evidence, an upper confidence limit of the frequency should be used in court.) If a particular DNA pattern occurred in 1 of 100 samples, the estimated frequency would be 1%, with an upper confidence limit of 4.7%. If the pattern occurred in 0 of 100 samples, the estimated frequency would be 0%, with an upper confidence limit of 3%. (The upper bound cited is the traditional 95% confidence limit, whose use implies that the true value has only a 5% chance of exceeding the upper bound.) Such estimates produced by straightforward counting have the virtue that they do not depend on theoretical assumptions, but simply on the samples having been randomly drawn from the appropriate population. However, such estimates do not take advantage of the full potential of the genetic approach.

In contrast, population frequencies often quoted for DNA typing analyses are based not on actual counting, but on theoretical models based on the principles of population genetics. Each matching allele is assumed to provide statistically independent evidence, and the frequencies of the individual alleles are multiplied together to calculate a frequency of the complete DNA pattern. Although a databank might contain only 500 people, multiplying the frequencies of enough separate events might result in an estimated frequency of their all occurring in a given person of 1 in a billion. Of course, the scientific validity of the multiplication rule depends on whether the events (i.e., the matches at each allele) are actually statistically independent.

Because it is impossible or impractical to draw a large enough population to test directly calculated frequencies of any particular DNA profile much below 1 in 1,000, there is not a sufficient body of empirical data on which to base a claim that such frequency calculations are reliable or valid. The assumption of independence must be strictly scrutinized and estimation procedures appropriately adjusted if possible. (The rarity of all the genotypes represented in the databank can be demonstrated by pairwise comparisons, however. Thus, in a recently reported analysis of the FBI databank, no exactly matching pairs were found in five-locus DNA profiles, and the closest match was a single three-locus match among 7.6 million pairwise comparisons.)

The multiplication rule has been routinely applied to blood-group frequencies in the forensic setting. However, that situation is substantially different. Because conventional genetic markers are only modestly polymorphic (with the exception of human leukocyte antigen, HLA, which usually cannot be typed in forensic specimens), the multilocus genotype frequencies are often about 1 in 100. Such estimates have been tested by simple empirical counting. Pairwise comparisons of allele frequencies have not revealed any correlation across loci. Hence, the multiplication rule does not appear to lead to the risk of extrapolating beyond the available data for conventional markers. But highly polymorphic DNA markers exceed the informative power of protein markers and so multiplication of their estimated frequencies leads to estimates that are far less than the reciprocal of the size of the databanks, i.e., 1/N, N being the number of entries in the databank.

The multiplication rule is based on the assumption that the population does not contain subpopulations with distinct allele frequencies—that each person's alleles constitute statistically independent random selections from a common gene pool. Under that assumption, the procedure for calculating the population frequency of a genotype is straightforward:

• Count the frequency of alleles. For each allele in the genotype, examine a random sample of the population and count the proportion of matching alleles—that is, alleles that would be declared to match according to the rule that is used for declaring matches in a forensic context.

• Calculate the frequency of the genotype at each locus. The frequency of a homozygous genotype a1/a1 is calculated to be p_{a1}^2, where p_{a1} denotes the frequency of allele a1. The frequency of a heterozygous genotype a1/a2 is calculated to be $2p_{a1}p_{a2}$, where p_{a1} and p_{a2} denote the frequencies of alleles a1 and a2. In both cases, the genotype frequency is calculated by multiplying the two allele frequencies, on the assumption that there is no statistical correlation between the allele inherited from one's father and the allele inherited from one's mother. When there is no correlation between the two parental alleles, the locus is said to be in Hardy-Weinberg equilibrium.

• Calculate the frequency of the complete multilocus genotype by multiplying the genotype frequencies at all the loci. As in the previous step, this calculation assumes that there is no correlation between the genotypes at the individual loci; the absence of such correlation is called linkage equilibrium. Suppose, for example, that a person has genotype a1/a2, b1/b2, c1/c1. If a random sample of the appropriate population shows that the frequencies of alleles a1, a2, b1, b2, and c1 are approximately 0.1, 0.2, 0.3, 0.1, and 0.2, respectively, then the population frequency of the genotype would be estimated to be [2(0.1)(0.2)][2(0.3)(0.1)][(0.2)(0.2)] = 0.000096, or about 1 in 10,417.

The validity of the multiplication rule depends on the assumption of absence of population substructure. Population substructure violates the assumption of statistical independence of alleles. In a population that contains groups each with different allele frequencies, the presence of one allele in a person's genotype can alter the statistical expectation of the other alleles in the genotype. For example, a person who has one allele that is common among Italians is more likely to be of Italian descent and is thus more likely to carry additional alleles that are common among Italians. The true genotype frequency is thus higher than would be predicted by applying the multiplication rule using the average frequency in the entire population.

To illustrate the problem with a hypothetical example, suppose that a particular allele at a VNTR locus has a 1% frequency in the general population, but a 20% frequency in a specific subgroup. The frequency of homozygotes for the allele would be calculated to be 1 in 10,000 according to the allele frequency determined by sampling the general population, but would actually be 1 in 25 for the subgroup. That is a hypothetical and extreme example, but illustrates the potential effect of demography on gene frequency estimation.

The key question underlying the use of the multiplication rule—i.e., whether actual populations have significant substructure for the loci used for forensic typing—has provoked considerable debate among population geneticists. Some have expressed serious concern about the possibility of significant substructure. They maintain that census categories—such as North American Caucasians, blacks, Hispanics, Asians, and Native Americans—are not homogeneous groups, but rather that each group is an admixture of subgroups with somewhat different allele frequencies. Allele frequencies have not yet been homogenized, because people tend to mate within their subgroups.

Those population geneticists also point out that, for any particular genetic marker, the actual degree of subpopulation differentiation cannot be predicted in advance, but must be determined empirically. Furthermore, they doubt that the presence of substructure can be detected by the application of statistical tests to data from large mixed populations. Population differentiation must be assessed through direct studies of allele frequencies in ethnic groups.

Other population geneticists, while recognizing the possibility or likelihood of population substructure, conclude that the evidence to date suggests only a minimal effect on estimates of genotype frequencies. Recent empirical studies concerning VNTR loci detected no deviation from independence within or across loci. Moreover, as pointed out earlier, pairwise comparisons of all five-locus DNA profiles in the FBI database showed no exact matches; the closest match was a single three-locus match among 7.6 million pairwise comparisons. Those studies are interpreted as indicating that multiplication of gene frequencies across loci does not lead to major inaccuracies in the calculation of genotype frequency—at least not for the specific polymorphic loci examined.

Although mindful of those opposing views, the committee has chosen to assume for the sake of discussion that population substructure may exist and to provide a method for estimating population frequencies in a manner that would adequately account for it. Our decision is based on four considerations:

1. It is possible to provide conservative estimates of population frequency, without giving up the inherent power of DNA typing.

2. It is appropriate to prefer somewhat conservative numbers for forensic DNA typing, especially because the statistical power lost in this way can often be recovered through typing of additional loci, where required.

3. It is important to have a general approach that is applicable to any loci used for forensic typing. Recent empirical studies pertain only to the population genetics of the VNTR loci in current use. However, we expect

forensic DNA typing to undergo much change over the next decade—including the introduction of different types of DNA polymorphisms, some of which might have different properties from the standpoint of population genetics.

4. It is desirable to provide a method for calculating population frequencies that is independent of the ethnic group of the subject.

The committee is aware of the need to account for possible population substructure, and it recommends the use of the ceiling principle. The multiplication rule will yield conservative estimates even for a substructured population, provided that the allele frequencies used in the calculation exceed the allele frequencies in any of the population subgroups. The ceiling principle involves two steps: (1) for each allele at each locus, determine a *ceiling frequency* that is an upper bound of the allele frequency that is independent of the ethnic background of a subject; and (2) to calculate a genotype frequency, apply the multiplication rule according to the ceiling allele frequencies.

To determine ceiling frequencies, the committee strongly recommends the following approach: (1) Draw random samples of 100 persons from each of 15-20 populations that represent groups relatively homogeneous genetically. (2) Take as the ceiling frequency the largest frequency in any of those populations or 5%, whichever is larger.

Use of the ceiling principle yields the same frequency of a given genotype, regardless of the suspect's ethnic background, because the reported frequency represents a maximum for any possible ethnic heritage. Accordingly, the ethnic background of the individual suspect should be ignored in estimating the likelihood of a random match. The calculation is fair to suspects, because the estimated probabilities are likely to be conservative in their incriminating power.

Some legal commentators have pointed out that frequencies should be based on the population of possible perpetrators, rather than on the population to which a particular suspect belongs. Although that argument is formally correct, practicalities often preclude use of that approach. Furthermore, the ceiling principle eliminates the need for investigating the perpetrator population, because it yields an upper bound to the frequency that would be obtained by that approach.

Although the ceiling principle is a conservative approach, we feel that it is appropriate. DNA typing is unique, in that the forensic analyst has an essentially unlimited ability to adduce additional evidence: whatever power is sacrificed by requiring conservative estimates can be regained by examining additional loci. (Although there might be some cases in which the DNA sample is insufficient to permit typing additional loci with RFLPs, this limitation is likely to disappear with the eventual use of PCR.)

That no evidence of population substructure is demonstrable with the markers tested so far cannot be taken to mean that such does not exist for other markers. Preservation of population DNA samples in the form of immortalized cell lines will ensure that DNA is available for determining population frequencies of any DNA pattern as new and better techniques become available, without the necessity of collecting fresh samples. It will also provide samples for standardization of methods across laboratories.

Because of the similarity in DNA patterns between relatives, databanks of DNA of convicted criminals have the ability to point not just to individuals but to entire families—including relatives who have committed no crime. Clearly, this raises serious issues of privacy and fairness. It is inappropriate, for reasons of privacy, to search databanks of DNA from convicted criminals in such a fashion. Such uses should be prevented both by limitations of the software for searching and by statutory guarantees of privacy.

The genetic correlation among relatives means that the probability that a forensic sample will match a relative of the person who left it is considerably greater than the probability that it will match a random person.

Especially for a technology with high discriminatory power, such as DNA typing, laboratory error rates must be continually estimated in blind proficiency testing and must be disclosed to juries.

Recommendations

- As a basis for the interpretation of the statistical significance of DNA typing results, the committee recommends that blood samples be obtained from 100 randomly selected persons in each of 15-20 relatively homogeneous populations; that the DNA in lymphocytes from these blood samples be used to determine the frequencies of alleles currently tested in forensic applications; and that the lymphocytes be "immortalized" and preserved as a reference standard for determination of allele frequencies in tests applied in different laboratories or developed in the future. The collection of samples and their study should be overseen by a National Committee on Forensic DNA Typing.
- The ceiling principle should be used in applying the multiplication rule for estimating the frequency of particular DNA profiles. For each allele in a person's DNA pattern, the highest allele frequency found in any of the 15-20 populations or 5% (whichever is larger) should be used.
- In the interval (which should be short) while the reference blood samples are being collected, the significance of the findings of multilocus DNA typing should be presented in two ways: (1) If no match is found with any sample in a total databank of N persons (as will usually be the case), that should be stated, thus indicating the rarity of a random match. (2) In applying the multiplication rule, the 95% upper confidence limit of the frequency of each allele should be calculated for separate U.S. "racial"

groups and the highest of these values or 10% (whichever is the larger) should be used. Data on at least three major "races" (e.g., Caucasians, blacks, Hispanics, Asians, and Native Americans) should be analyzed.

- Any population databank used to support DNA typing should be openly available for scientific inspection by parties to a legal case and by the scientific community.
- Laboratory error rates should be measured with appropriate proficiency tests and should play a role in the interpretation of results of forensic DNA typing.

STANDARDS

Critics and supporters of the forensic uses of DNA typing agree that there is a lack of standardization of practices and a lack of uniformly accepted methods for quality assurance. The deficiencies are due largely to the rapid emergence of DNA typing and its introduction in the United States through the private sector.

As the technology developed in the United States, private laboratories using widely differing methods (single-locus RFLP, multilocus RFLP, and PCR) began to offer their services to law-enforcement agencies. During the same period, the FBI was developing its own RFLP method, with a different restriction enzyme and different single-locus probes. The FBI's method has become the one most widely used in public forensic-science laboratories. Each method has its own advantages and disadvantages, databanks, molecular-weight markers, match criteria, and reporting methods.

Regardless of the causes, practices in DNA typing vary, and so do the educational backgrounds, training, and experience of the scientists and technicians who perform the tests, the internal and external proficiency testing conducted, the interpretation of results, and approaches to quality assurance.

It is not uncommon for an emerging technology to go without regulation until its importance and applicability are established. Indeed, the development of DNA typing technology has occurred without regulation of laboratories and their practices, public or private. The committee recognizes that standardization of practices in forensic laboratories in general is more problematic than in other laboratory settings; stated succinctly, forensic scientists have little or no control over the nature, condition, form, or amount of sample with which they must work. But it is now clear that DNA typing methods are a most powerful adjunct to forensic science for personal identification and have immense benefit to the public—so powerful, so complex, and so important that some degree of standardization of laboratory procedures is necessary to assure the courts of high-quality results. DNA typing is capable, in principle, of an extremely low inherent rate of false results, so the risk of error will come from poor laboratory practice or poor sample handling and labeling; and, because DNA typing is technical, a jury requires the assurance of laboratory competence in test results.

At issue, then, is how to achieve standardization of DNA typing laboratories in such a manner as to assure the courts and the public that results of DNA typing by a given laboratory are reliable, reproducible, and accurate.

Quality assurance can best be described as a documented system of activities or processes for the effective monitoring and verification of the quality of a work product (in this case, laboratory results). A comprehensive quality-assurance program must include elements that address education, training, and certification of personnel; specification and calibration of equipment and reagents; documentation and validation of analytical methods; use of appropriate standards and controls; sample handling procedures; proficiency testing; data interpretation and reporting; internal and external audits of all the above; and corrective actions to address deficiencies and weigh their importance for laboratory competence.

Recommendations

Although standardization of forensic practice is difficult because of the nature of the samples, DNA typing is such a powerful and complex technology that some degree of standardization is necessary to ensure high standards.

• Each forensic-science laboratory engaged in DNA typing must have a formal, detailed quality-assurance and quality-control program to monitor work, on both an individual and a laboratory-wide basis.

• The Technical Working Group on DNA Analysis and Methods (TWGDAM) guidelines for a quality-assurance program for DNA RFLP analysis are an excellent starting point for a quality-assurance program, which should be supplemented by the additional technical recommendations of this committee.

• The TWGDAM group should continue to function, playing a role complementary to that of the National Committee on Forensic DNA Typing (NCFDT). To increase its effectiveness, TWGDAM should include additional technical experts from outside the forensic community who are not closely tied to any forensic laboratory.

• Quality-assurance programs in individual laboratories alone are insufficient to ensure high standards. External mechanisms are needed, to ensure adherence to the practices of quality assurance. Potential mechanisms include individual certification, laboratory accreditation, and state or federal regulation.

• One of the best guarantees of high quality is the presence of an active professional-organization committee that is able to enforce standards. Although professional societies in forensic science have historically not played an active role, the American Society of Crime Laboratory Directors (ASCLD) and the American Society of Crime Laboratory Directors-Laboratory Accreditation Board (ASCLD-LAB) recently have shown substantial interest in enforcing quality by expanding the ASCLD-LAB accreditation program to include mandatory proficiency testing. ASCLD-LAB must demonstrate that it will actively discharge this role.

• Because private professional organizations lack the regulatory authority to require accreditation, further means are needed to ensure compliance with appropriate standards.

• Courts should require that laboratories providing DNA typing evidence have proper accreditation for each DNA typing method used. Any laboratory that is not formally accredited and that provides evidence to the courts—e.g., a nonforensic laboratory repeating the analysis of a forensic laboratory—should be expected to demonstrate that it is operating at the same level of standards as accredited laboratories.

• Establishing mandatory accreditation should be a responsibility of the Department of Health and Human Services (DHHS), in consultation with the Department of Justice (DOJ). DHHS is the appropriate agency, because it has extensive experience in the regulation of clinical laboratories through programs under the Clinical Laboratory Improvement Act and has extensive expertise in molecular genetics through the National Institutes of Health. DOJ must be involved, because the task is important for law enforcement.

• The National Institute of Justice (NIJ) does not appear to receive adequate funds to support proper education, training, and research in the field of forensic DNA typing. The level of funding should be re-evaluated and increased appropriately.

DATABANKS AND PRIVACY OF INFORMATION

DNA typing in the criminal-justice system has so far been used primarily for direct comparison of DNA profiles of evidence samples with profiles of samples from suspects. However, that application constitutes only the tip of the iceberg of potential law-enforcement applications. If DNA profiles of samples from a population were stored in computer databanks (databases), DNA typing could be applied in crimes without suspects. Investigators could compare DNA profiles of biological evidence samples with profiles in a databank to search for suspects.

In many respects, the situation is analogous to that of latent fingerprints. Originally, latent fingerprints were used for comparing crime-scene evidence with suspects. With the development of the Automated Fingerprint Identification Systems (AFIS) in the last decade, the investigative use of fingerprints has dramatically expanded. Forensic scientists can enter an unidentified latent-fingerprint pattern into an automated system and within minutes compare it with millions of person's patterns contained in a computer file. In its short history, automated fingerprint analysis has been credited with solving tens of thousands of crimes.

The computer technology required for an automated fingerprint identification system is sophisticated and complex. Fingerprints are complicated geometric patterns, and the computer must store, recognize, and search for complex and variable patterns of ridges and minutiae in the millions of prints on file. Several commercially available but expensive computer systems are in use around the world. In contrast, the computer technology required for DNA databanks is relatively simple. Because DNA profiles can be reduced to a list of genetic types (hence, a list of numbers), DNA profile repositories can use relatively simple and inexpensive software and hardware. Consequently, computer requirements should not pose a serious problem in the development of DNA profile databanks.

Confidentiality and security of DNA-related information are especially important and difficult issues, because we are in the midst of two extraordinary technological revolutions that show no signs of abating: in molecular biology, which is yielding an explosion of information about human genetics, and in computer technology, which is moving toward national and international networks connecting growing information resources.

Even simple information about identity requires confidentiality. Just as fingerprint files can be misused, DNA profile information could be misused to search and correlate criminal-record databanks or medical-record databanks. Computer storage of information increases the possibilities for misuse. For example, addresses, telephone numbers, social security numbers, credit ratings, range of incomes, demographic categories, and information on hobbies are currently available for many of our citizens in various distributed computerized data sources. Such data can be obtained directly through access to specific sources, such as credit-rating services, or through statistical disclosure, which refers to the ability of a user to derive an estimate of a desired statistic or feature from a databank or a collection of databanks. Disclosure can be achieved through one query or a series of queries to one or more databanks. With DNA information, queries might be directed at obtaining numerical estimates of values or at deducing the state of an attribute of an individual through a series of Boolean (yes-no) queries to multiple distributed databanks.

Several private laboratories already offer a DNA-banking service (sample storage in freezers) to physicians, genetic counselors, and, in some cases, anyone who pays for the service. Typically, such information as name, address, birth date, diagnosis, family history, physician's name and address, and genetic counselor's name and address is stored with samples.

That information is useful for local, independent bookkeeping and record management. But it is also ripe for statistical or correlative disclosure. Just the existence in a databank of a sample from a person, independent of any DNA-related information, may be prejudicial to the person. In some laboratories, the donor cannot legally prevent outsiders' access to the samples, but can request its withdrawal. A request for withdrawal might take a month or more to process. In most cases, only physicians with signed permission of the donor have access to samples, but typically no safeguards are taken to verify individual requests independently. That is not to say that the laboratories intend to violate donors' rights; they are simply offering a service for which there is a recognized market and attempting to provide services as well as they can.

Recommendations

• In the future, if pilot studies confirm its value, a national DNA profile databank should be created that contains information on felons convicted of particular violent crimes. Among crimes with high rates of recidivism, the case is strongest for rape, because perpetrators typically leave biological evidence (semen) that could allow them to be identified. Rape is the crime for which the databank will be of primary use. The case is somewhat weaker for violent offenders who are most likely to commit homicide as a recidivist offense, because killers leave biological evidence only in a minority of cases.

• The databank should also contain DNA profiles of unidentified persons made from biological samples found at crime scenes. These would be samples known to be of human origin, but not matched with any known persons.

• Databanks containing DNA profiles of members of the general population (as exist for ordinary fingerprints for identification purposes) are not appropriate, for reasons of both privacy and economics.

• DNA profile databanks should be accessible only to legally authorized persons and should be stored in a secure information resource.

• Legal policy concerning access and use of both DNA samples and DNA databank information should be established before widespread proliferation of samples and information repositories. Interim protection and sanctions against misuse and abuse of information derived from DNA typing should be established immediately. Policies should explicitly define authorized uses and should provide for criminal penalties for abuses.

• Although the committee endorses the concept of a limited national DNA profile databank, it doubts that existing RFLP-based technology provides an appropriate wise long-term foundation for such a databank. We expect current methods to be replaced soon with techniques that are simpler, easier to automate, and less expensive—but incompatible with existing DNA profiles. Accordingly, the committee does not recommend establishing a comprehensive DNA profile databank yet.

- For the short term, we recommend the establishment of pilot projects that involve prototype databanks based on RFLP technology and consisting primarily of profiles of violent sex offenders. Such pilot projects could be worthwhile for identifying problems and issues in the creation of databanks. However, in the intermediate term, more efficient methods will replace the current one, and the forensic community should not allow itself to become locked into an outdated method.
- State and federal laboratories, which have a long tradition and much experience in the management of other types of basic evidence, should be given primary responsibility, authority, and additional resources to handle forensic DNA testing and all the associated sample-handling and data-handling requirements.
- Private-sector firms should not be discouraged from continuing to prepare and analyze DNA samples for specific cases or for databank samples, but they must be held accountable for misuse and abuse to the same extent as government-funded laboratories and government authorities.

DNA INFORMATION IN THE LEGAL SYSTEM

To produce biological evidence that is admissible in court in criminal cases, forensic investigators must be well trained in the collection and handling of biological samples for DNA analysis. They should take care to minimize the risk of contamination and ensure that possible sources of DNA are well preserved and properly identified. As in any forensic work, they must attend to the essentials of preserving specimens, labeling, and the chain of custody and must observe constitutional and statutory requirements that regulate the collection and handling of samples. The Fourth Amendment provides much of the legal framework for the gathering of DNA samples from suspects or private places, and court orders are sometimes needed in this connection.

In civil (noncriminal) cases—such as paternity, custody, and proof-of-death cases—the standards for admissibility must also be high, because DNA evidence might be dispositive. The relevant federal rules (Rules 403 and 702-706) and most state rules of evidence do not distinguish between civil and criminal cases in determining the admissibility of scientific data. In a civil case, however, if the results of a DNA analysis are not conclusive, it will usually be possible to obtain new samples for study.

The advent of DNA typing technology raises two key issues for judges: determining *admissibility* and explaining to jurors the appropriate standards for *weighing* evidence. A host of subsidiary questions with respect to how expert evidence should be handled before and during a trial to ensure prompt and effective adjudication apply to all evidence and all experts.

In the United States, there are two main tests for admissibility of scientific information through experts. One is the Frye test, enunciated in *Frye v. United States*. The other is a "helpfulness" standard found in the Federal Rules of Evidence and many of its state counterparts. In addition, several states have recently enacted laws that essentially mandate the admission of DNA typing evidence.

The test for the admissibility of novel scientific evidence enunciated in *Frye v. United States* is still probably the most frequently invoked test in American case law. A majority of states profess adherence to the *Frye* rule, although a growing minority have adopted variations on the helpfulness standard suggested by the Federal Rules of Evidence.

Frye predicates the admissibility of novel scientific evidence on its general acceptance in a particular scientific field: "While courts will go a long way in admitting expert testimony deduced from a well-recognized scientific principle or discovery, the thing from which the deduction is made must be sufficiently established to have gained general acceptance in the particular field in which it belongs." Thus, admissibility depends on the quality of the science underlying the evidence, as determined by scientists themselves. Theoretically, the court's role in this preliminary determination is narrow: it should conduct a hearing to determine whether the scientific theory underlying the evidence is generally accepted in the relevant scientific community and whether the specific techniques used are reliable for their intended purpose.

In practice, the court is much more involved. The court must determine the scientific fields from which experts should be drawn. Complexities arise with DNA typing, because the full typing process rests on theories and findings that pertain to various scientific fields. For example, the underlying theory of detecting polymorphisms is accepted by human geneticists and molecular biologists, but population geneticists and other statisticians might differ as to the appropriate method for determining the population frequency of a genotype in the general population or in a particular geographic, ethnic, or other group. The courts often let experts on a process, such as DNA typing, testify to the various scientific theories and assumptions on which the process rests, even though the experts' knowledge of some of the underlying theories is likely to be at best that of a generalist, rather than a specialist.

The *Frye* test sometimes prevents scientific evidence from being presented to a jury unless it has sufficient history to be accepted by some subspecialty of science. Under *Frye*, potentially helpful evidence may be excluded until consensus has developed. By 1991, DNA evidence had been considered in hundreds of *Frye* hearings involving felony prosecutions in more than 40 states. The overwhelming majority of trial courts ruled that such evidence was admissible, but there have been some important exceptions.

In determining admissibility according to the helpfulness standard under the Federal Rules of Evidence, without specifically repudiating the *Frye* rule, a court can adopt a more flexible approach. Rule 702 states that, "if scientific, technical or other specialized knowledge will assist the trier of fact to understand the evidence or to determine a fact in issue, a witness qualified as an expert by knowledge, skill, experience, training, or education, may testify thereto in the form of an opinion or otherwise."

Rule 702 should be read with Rule 403, which requires the court to determine the admissibility of evidence by balancing its probative force against its potential for misapplication by the jury. In determining admissibility, the court should consider the soundness and reliability of the process or technique used in generating evidence; the possibility that admitting the evidence would overwhelm, confuse, or mislead the jury; and the proffered connection between the scientific research or test result to be presented and particular disputed factual issues in the case.

The federal rule, as interpreted by some courts, encompasses *Frye* by making general acceptance of scientific principles by experts a factor, and in some cases a decisive factor, in determining probative force. A court can also consider the qualifications of experts testifying about the new scientific principle, the use to which the technique based on the principle has been put, the technique's potential for error, the existence of specialized literature discussing the technique, and its novelty.

With the helpfulness approach, the court should also consider factors that might prejudice the jury. One of the most serious concerns about scientific evidence, novel or not, is that it possesses an aura of infallibility that could overwhelm a jury's critical faculties. The likelihood that the jury would abdicate its role as critical fact-finder is believed by some to be greater if the science underlying an expert's conclusion is beyond its intellectual grasp. The jury might feel compelled to accept or reject a conclusion absolutely or to ignore evidence altogether. However, some experience indicates that jurors tend not to be overwhelmed by scientific proof and that they prefer experiential data based on traditional forms of evidence. Moreover, the presence of opposing experts might prevent a jury from being unduly impressed with one expert or the other. Conversely, the absence of an opposing expert might cause a jury to give too much weight to expert testimony, on the grounds that, if the science were truly controversial, it would have heard the opposing view. Nevertheless, if the scientific evidence is valid, the solution to those possible problems is not to exclude the evidence, but to ensure through instructions and testimony that the jury is equipped to consider rationally whatever evidence is presented.

In determining admissibility with the helpfulness approach, the court should consider a number of factors in addition to reliability. First is the significance of the issue to which the evidence is directed. If the issue is tangential to the case, the court should be more reluctant to allow a time-consuming presentation of scientific evidence that might itself confuse the jury. Second, the availability and sufficiency of other evidence might make expert testimony about DNA superfluous. And third, the court should be mindful of the need to instruct and advise the jury so as to eliminate the risk of prejudice.

Recommendations

• Courts should take judicial notice of three scientific underpinnings of DNA typing:

—The study of DNA polymorphisms can, in principle, provide a reliable method for comparing samples.
—Each person's DNA is unique (except that of identical twins), although the actual discriminatory power of any particular DNA test will depend on the sites of DNA variation examined.
—The current laboratory procedure for detecting DNA variation (specifically, single-locus probes analyzed on Southern blots without evidence of band shifting) is fundamentally sound, although the validity of any particular implementation of the basic procedure will depend on proper characterization of the reproducibility of the system (e.g., measurement variation) and inclusion of all necessary scientific controls.

• The adequacy of the method used to acquire and analyze samples in a given case bears on the admissibility of the evidence and should, unless stipulated by opposing parties, be adjudicated case by case. In this adjudication, the accreditation and certification status of the laboratory performing the analysis should be taken into account.

• Because of the potential power of DNA evidence, authorities should make funds available to pay for expert witnesses, and the appropriate parties must be informed of the use of DNA evidence as soon as possible.

• DNA samples (and evidence likely to contain DNA) should be preserved whenever that is possible.

• All data and laboratory records generated by analysis of DNA samples should be made freely available to all parties. Such access is essential for evaluating the analysis.

• Protective orders should be used only to protect the privacy of individuals.

DNA TYPING AND SOCIETY

The introduction of any new technology is likely to raise concerns about its impact on society. Financial costs, potential harm to the interests of individuals, and threats to liberty or privacy are only a few of the worries typically voiced when a new technology is on the horizon. DNA typing technology has the potential for uncovering and revealing a great deal of information that most people consider to be intensely private. Examples might be the presence of genes involved in known genetic disorders or genes that have been linked to a heightened risk of particular major diseases in some populations.

Although DNA technology involves new scientific techniques for identifying or excluding people, the techniques are extensions and analogues of techniques long used in forensic science, such as serological and fingerprint examinations. Ethical questions can be raised about other aspects of this new technology, but the committee does not see it as violating a fundamental ethical principle.

A new practice or technology can be subjected to further ethical analysis by using two leading ethical perspectives. The first examines the action or practice in terms of the rights of people who are affected; the second explores the potential positive and negative consequences (nonmonetary costs and benefits) of the action or practice, in an attempt to determine whether the potential good consequences outweigh the bad.

Two main questions can be asked about moral rights: Does the use of DNA technology give rise to any new rights not already recognized? Does the use of DNA technology enhance, endanger, or diminish the rights of anyone who becomes involved in legal proceedings? In answer to the first question, it is hard to think of any new rights not already recognized that come into play with the introduction of DNA technology into forensic science. The answer to the second question requires a specification of the classes of people whose rights might be affected and what those rights might be.

Concerns about intrusions into privacy and breaches of confidentiality regarding the use of DNA technology in such enterprises as gene mapping are frequently voiced, and they are legitimate ethical worries. The concerns are pertinent to the role of DNA technology in forensic science, as well as to its widespread use for other purposes and in other social contexts. A potential problem related to the confidentiality of any information obtained is the safeguarding of the information and the prevention of its unauthorized release or dissemination; that can also be classified under the heading of abuse and misuse, as well as seen as a violation of individual rights in the forensic context.

Another factor to be weighed in a consequentialist ethical analysis is whose interests are to count and whether some people's interests should be given greater weight than others'. For example, there are the interests of the accused, the interests of victims of crime or their families in apprehending and convicting perpetrators, and the interests of society. Whether the interests of society in seeing that justice is done should count as much as the interests of the accused or the victim is open to question.

A major issue is the preservation of confidentiality of information obtained with DNA technology in the forensic context. When databanks are established in such a way that state and federal law-enforcement authorities can gain access to DNA profiles, not only of persons convicted of violent crimes but of others as well, there is a serious potential for abuse of confidential information. The victims of many crimes in urban areas are relatives or neighbors of the perpetrators, and these victims might themselves be former or future perpetrators. There is greater likelihood that DNA information on minority-group members, such as blacks and Hispanics, will be stored or accessed. However, it is important to note that use of the ceiling principle removes the necessity to categorize criminals (or defendants in general) by ethnic group for the purposes of DNA testing and storage of information in databanks.

The introduction of a powerful new technology is likely to set up expectations that might be unwarranted or unrealistic in practice. Various expectations regarding DNA typing technology are likely to be raised in the minds of jurors and others in the forensic setting. For example, public perception of the accuracy and efficacy of DNA typing might well put pressure on prosecutors to obtain DNA evidence whenever appropriate samples are available. As the use of the technology becomes widely publicized, juries will come to expect it, just as they now expect fingerprint evidence.

Two aspects of DNA typing technology contribute to the likelihood of its raising inappropriate expectations in the minds of jurors. The first is a jury's perception of an extraordinarily high probability of enabling a definitive identification of a criminal suspect; the second is the scientific complexity of the technology, which results in laypersons' inadequate understanding of its capabilities and failings. Taken together, those two aspects can lead to a jury's ignoring other forensic evidence that it should be considering.

As large felon databanks are created, the forensic community could well place more reliance on DNA evidence, and a possible consequence is the underplaying of other forensic evidence. Unwarranted expectations about the power of DNA technology might result in the neglect of relevant evidence.

The need for international cooperation in law enforcement calls for appropriate scientific and technical exchange among nations. As in other areas of science and technology, dissemination of information about DNA typing and training programs for personnel likely to use the technology should be encouraged. It is desirable that all nations that will collaborate in law-enforcement activities have similar standards and practices, so efforts should be furthered to exchange scientific knowledge and expertise regarding DNA technology in forensic science.

Recommendations

- In the forensic context as in the medical setting, DNA information is personal, and a person's privacy and need for confidentiality should be respected. The release of DNA information on a criminal population without the subjects' permission for purposes other than law enforcement should be considered a misuse of the information, and legal sanctions should be established to deter the unauthorized dissemination or procurement of DNA information that was obtained for forensic purposes.
- Prosecutors and defense counsel should not oversell DNA evidence. Presentations that suggest to a judge or jury that DNA typing is infallible are rarely justified and should be avoided.
- Mechanisms should be established to ensure accountability of laboratories and personnel involved in DNA typing and to make appropriate public scrutiny possible.
- Organizations that conduct accreditation or regulation of DNA technology for forensic purposes should not be subject to the influence of private companies, public laboratories, or other organizations actually engaged in laboratory work.
- Private laboratories used for testing should not be permitted to withhold information from defendants on the grounds that trade secrets are involved.
- The same standards and peer-review processes used to evaluate advances in biomedical science and technology should be used to evaluate forensic DNA methods and techniques.
- Efforts at international cooperation should be furthered, in order to ensure uniform international standards and the fullest possible exchange of scientific knowledge and technical expertise.

1

Introduction

BACKGROUND

Characterization, or "typing," of blood, semen, and other body fluids has been used for forensic purposes for more than 50 years.[1] It began with blood groups, such as those of the ABO system, and later was extended to serum proteins and red-cell enzymes and in some forensic applications, particularly paternity testing, to human leukocyte antigens (HLA), which are associated with tissue types. The genetically determined person-to-person variation revealed by such typing was used mainly to include or exclude suspects, that is, to determine whether a person showed a combination of genetically determined characteristics consistent with having been the source of an evidence sample in a criminal case or having been the father of a child in a paternity case. Except when HLA testing was used, the chance that a randomly chosen person would be excluded by the tests was about 98%; that left a 2% chance that the test would "include" an innocent person.

In the last decade, methods have become available for deoxyribonucleic acid (DNA) typing, that is, for showing distinguishing differences in the genetic material itself. Advances in DNA technology in the 1970s paved the way for the detection of variation (polymorphism) in specific DNA sequences and shifted the study of human variation from the protein products of DNA to DNA itself. By analyzing a sufficient number of regions of DNA that show much person-to-person variability, one can reduce the probability of a chance match (inclusion) of two persons to an extremely low level. Indeed, the probability can, in principle, be made so low that DNA typing becomes not simply a method for exclusion or inclusion, but a means of absolute identification.

The potential applicability of DNA typing to forensic samples was demonstrated during the mid-1980s by laboratories in the United Kingdom, United States, and Canada. Their work established that DNA was present in forensic samples in sufficient quantity for testing (see Table 1.1) and that it survived in a state that allowed it to be typed. In the publications in 1985 by Jeffreys and colleagues,[2,3] the term "DNA fingerprint" carried the connotation of absolute identification. The mass-media coverage that accompanied the publications fixed in the general public's minds the idea that DNA typing could be used for absolute identification. Thus, the traditional forensic paradigm of genetic testing as a tool for exclusion was in a linguistic stroke changed to a paradigm of identification. (See Box 1 for a contrasting of dermatoglyphic fingerprints with "DNA fingerprints.")

Forensic DNA typing, first used in casework in 1985 in the United Kingdom, was initiated in the United States in late 1986 by commercial laboratories and in 1988 by the Federal Bureau of Investigation (FBI) and is now being used by dozens of state and local crime laboratories. Because of its great potential benefits for criminal and civil justice, but also because of the possibilities for its misuse or abuse, forensic DNA typing has been subjected to special scrutiny. Important questions have been asked about reliability, validity, and confidentiality:[4-6]

TABLE 1.1 DNA Content of Biological Samples

Type of Sample	Amount of DNA[a]
Blood	20,000-40,000 ng/ml
stain 1 cm^2 in area	ca. 200 ng
stain 1 mm^2 in area	ca. 2 ng
Semen	150,000-300,000 ng/ml
postcoital vaginal swab	0-3,000 ng
Hair:	
plucked	1-750 ng/hair
shed	1-12 ng/hair
Saliva	1,000-10,000 ng/ml
Urine	1-20 ng/ml

[a]The amount of DNA is given in nanograms (ng); 1 ng = one-billionth of a gram (10^{-9} g).

FINGERPRINTS IN PERSONAL IDENTIFICATION: DIFFERENCES BETWEEN DNA TYPING AND DERMATOGLYPHICS

Because of the central role of dermatoglyphic fingerprints in human identification, arising out of the personal uniqueness of the patterns, it is useful to compare and contrast traditional fingerprints with "DNA fingerprints."

Fingerprints were described as an individualizing characteristic as early as in 1892. The use of fingerprints in forensic science (and in relation to chromosomal abnormalities, such as Down syndrome, and other clinical disorders) was developed empirically without reference to the specific genetic basis of the patterns. Ridge count is a polygenic or multifactorial trait. The close correlation for ridge counts with that expected for an almost exclusively genetic trait, when "identical" and fraternal twins and other relatives are compared, supports polygenic inheritance.

In the forensic application, minutiae in the fingerprint patterns, not ridge counts, are used for personal identification. The minutiae result from random nongenetic events during embryonic development of the fingerpads. As a consequence, the patterns even of "identical" twins are distinguishable. Indeed, it appears that the fingerprint pattern of each human being is unique.

The distinction between the two types of fingerprints is illustrated by prints from "identical" twins shown here. The dermatoglyphic fingerprints shown in Figure B-a (Twin A) and Figure B-b (Twin B) are distinguished by the patterns of loops and whorls.

Finger	Twin A	Twin B
Right Hand		
Thumb (1)	Whorl	Whorl
2	Whorl	Central pocket whorl
3	Ulnar loop	Whorl
4	Whorl	Whorl
5	Ulnar loop	Ulnar loop
Left Hand		
Thumb (1)	Whorl	Whorl
2	Whorl	Ulnar loop
3	Ulnar loop	Ulnar loop
4	Whorl	Ulnar loop
5	Ulnar loop	Ulnar loop

However, typing of DNA from the blood of these twins in three laboratories showed a match for all tests. One laboratory, testing for variation in four chromosomes, estimated the population frequency of

continued on next page

continued from page 29

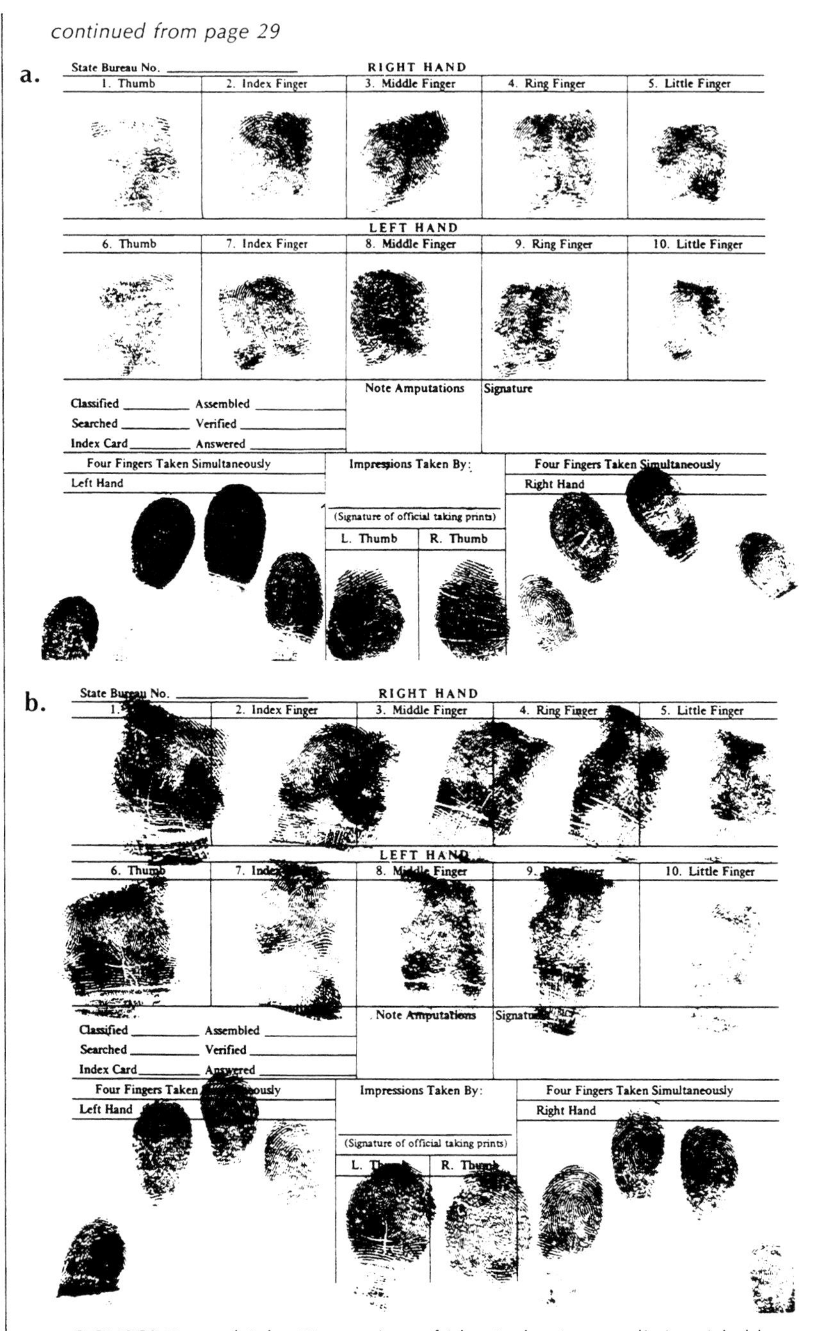

a.

State Bureau No. ____________

RIGHT HAND

1. Thumb	2. Index Finger	3. Middle Finger	4. Ring Finger	5. Little Finger

LEFT HAND

6. Thumb	7. Index Finger	8. Middle Finger	9. Ring Finger	10. Little Finger

Classified ________ Assembled ________
Searched ________ Verified ________
Index Card ________ Answered ________

Note Amputations

Signature

Four Fingers Taken Simultaneously — Left Hand

Impressions Taken By:

(Signature of official taking prints)

L. Thumb | R. Thumb

Four Fingers Taken Simultaneously — Right Hand

b.

State Bureau No. ____________

RIGHT HAND

1.	2. Index Finger	3. Middle Finger	4. Ring Finger	5. Little Finger

LEFT HAND

6. Thumb	7. [illegible]	8. [illegible] Finger	9. [illegible]	10. Little Finger

Classified ________ Assembled ________
Searched ________ Verified ________
Index Card ________ Answered ________

Note Amputations

Signature

Four Fingers Taken [illegible] — Left Hand

Impressions Taken By:

(Signature of official taking prints)

L. | R.

Four Fingers Taken Simultaneously — Right Hand

FIGURES B-a and B-b Fingerprints of identical twins are distinguishable.

the particular DNA patterns to be 1 in about 700,000. A second laboratory used four other probes and estimated the chance of a random match as 1 in about 1.8 million. For illustrative purposes, the patterns obtained with a "cocktail" of four single-locus probes are shown in Figure B-c. Twin A gave sample B, twin B gave sample E, and samples A, C, and D were from unrelated males of the same ethnic group. All five samples were submitted and tested in a blind manner. The other lanes show controls. (Courtesy of Robin W. Cotton and Matthew John McCoy, Cellmark Diagnostics.)

The uniqueness of the "DNA fingerprint" is based on genetic variation, whereas that of the dermatoglyphic fingerprint is based largely on nongenetic variation.

c.

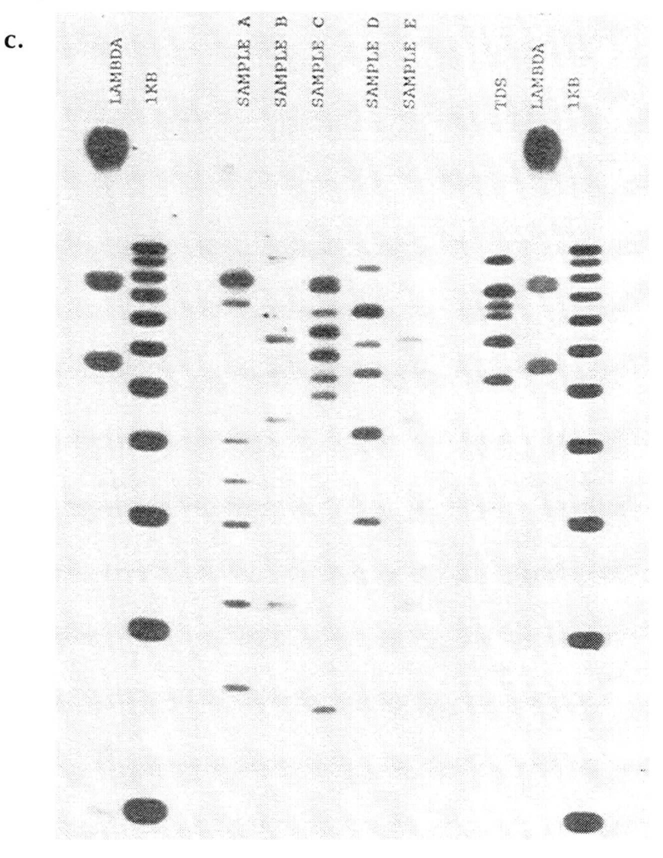

FIGURE B-c DNA types of identical twins are indistinguishable. Samples in lines A-E are from different white males; samples B and E are from identical twins.

- Criticisms were raised concerning the reliability of the technical methods, including the criteria for identifying DNA patterns and declaring matches, as well as quality control.
- Questions were raised about the validity of estimates of probability of random inclusion that were being presented in courts. Were the individual components of a specific DNA pattern statistically independent, so that it was proper to multiply their frequencies together in calculating the chance of a match? What population databanks were appropriate?
- Because DNA can be used to derive medical and other personal information, questions of confidentiality and privacy have assumed greater importance in DNA typing than in the use of non-DNA tests.

By the summer of 1989, because of questions concerning DNA typing raised in connection with some well-publicized criminal cases, the scientific and legal communities had called for an examination of the issues by the National Research Council of the National Academy of Sciences.[5,7,8] As a response, the Committee on DNA Technology in Forensic Science was formed, and its first meeting was held in January 1990. The committee was to address the general applicability and appropriateness of the use of DNA technology in forensic science, the need for standards in data collection and analysis, the need for advances in technology, management of DNA typing data, and legal, societal, and ethical issues surrounding DNA typing.

GENETIC BASIS OF DNA TYPING

Genetics is the science of biological variation. The fundamental basis of genetics and the essence of Mendel's discovery in 1865 is that inheritance is particulate and that the inherited factors (genes) that determine visible traits exist in pairs of alleles (i.e., alternative forms of a gene at a given site)—one on a chromosome inherited from the father and one on a chromosome from the mother. Chromosomes that contain genes are threadlike or rodlike structures in the cell nucleus. An organism's particular combination of alleles is referred to as the organism's genotype; the collection of traits resulting therefrom is referred to as the organism's phenotype. Most markers (i.e., identifiable physical locations on a chromosome) used in forensic DNA typing are not parts of expressed genes (i.e., genes that code for products like proteins); they are in noncoding portions of DNA. Hence, they are not associated with a phenotype.

A trait that differs among individuals is referred to as a polymorphism.[9] In DNA typing, that term is used interchangeably with "variation." The variations in blood groups, serum protein types, and HLA tissue types used for forensic testing in the pre-DNA era were polymorphisms in the protein product; these proteins contain variations that reflect variations in DNA. But DNA technology makes it possible to study the variations directly.

Structure and Function of DNA

A human has 22 pairs of nonsex chromosomes (autosomes) and two sex chromosomes—two X chromosomes in a female or an X chromosome and a Y chromosome in a male. Each autosome or X or Y chromosome is composed of a long DNA molecule constructed as a double helix (Figure 1-1). Each component strand of the double helix is a chain of nucleotides of four types designated by the names of the bases adenine (A), cytosine (C), guanine (G), and thymine (T). The nucleotides bond, A to T and C to G, between the two strands of the helix like the rungs of a ladder or, better, the steps in a spiral staircase. A pair of complementary nucleotides (or bases)—A-T, G-C, T-A, or C-G—is called a basepair (bp). DNA replication, which takes place in association with cell division, involves the separation of the two strands of the double helix and the synthesis of a new strand of nucleotides complementary to each strand.

Genes are segments of the DNA molecule. They constitute the blueprint for the structure of proteins of various types that are responsible for the makeup and function of cells and the body as a whole. A human has 50,000-100,000 genes, each occurring in every nucleated body cell. Chromosome 1, the largest, might, for example, have about 5,000 genes spaced at intervals along the DNA molecule that it consists of.

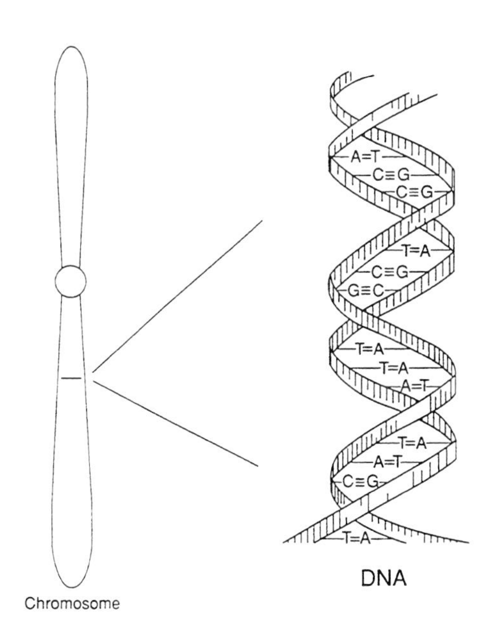

FIGURE 1-1 Diagram of the double-helical structure of DNA in a chromosome. The line shown in the chromosome is expanded to show the DNA structure.

Individual Variation in DNA

DNA technology has revealed variations in the genome, the total genetic makeup of the members of a species: single-nucleotide differences, deletions, and insertions. In noncoding regions of DNA, which are less constrained by forces of selection, it is estimated that at least one nucleotide per 300-1,000, on the average, varies between two people.[10] The nucleotide difference might change the recognition site for a particular site-specific endonuclease (restriction enzyme) so as to keep the DNA from being cut at that site by that enzyme. For example, in Figure 1-2 note that a single nucleotide change from C to T has eliminated a restriction enzyme cutting site. In addition, some regions of DNA contain repetitive units, multiple identical strings of nucleotides arranged in tandem. In VNTRs (variable

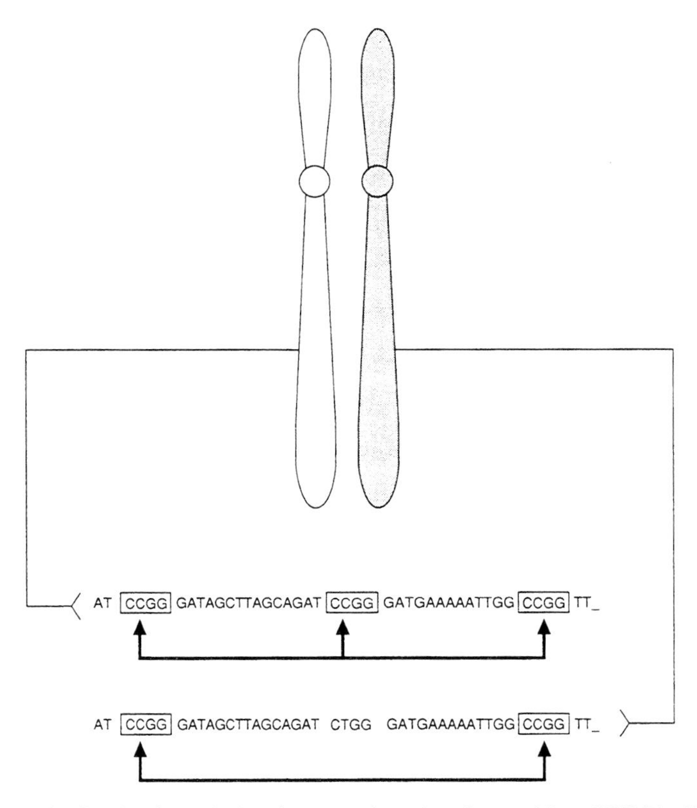

FIGURE 1-2 Basis of restriction fragment length polymorphism (RFLP). Diagram indicates heterozygosity at a "restriction site": chromosome on left has sequence CCGG that is recognized and cut by specific restriction enzyme, whereas chromosome on right has one-nucleotide difference that results in sequence CTGG, which is not recognized and cut by enzyme.

number tandem repeats), the number of repetitions of a sequence can vary from person to person. VNTRs are a leading form of variation used currently in forensic DNA typing. The repeating unit can be as small as a dinucleotide—e.g., the $(TG)_n$ polymorphism—or as large as 30, or even more, nucleotides. Tandem repeats are not limited to noncoding segments of DNA, although they are found less frequently in coding segments.

The two main types of variation—single-nucleotide differences and VNTRs—are both potentially recognizable by change in the lengths of fragments that result when DNA is cut with a restriction enzyme. Variation in the lengths of fragments can result from a change in the cluster of four, five, or six nucleotides that is the specific cutting site of the particular restriction enzyme (Figure 1-2). Or the variation can result, not from a change in the cutting site of the enzyme, but from the existence of different numbers of tandem repeats between two cutting sites. Figure 1-3 diagrams the major characteristics of the two forms of variation and their use in DNA typing.

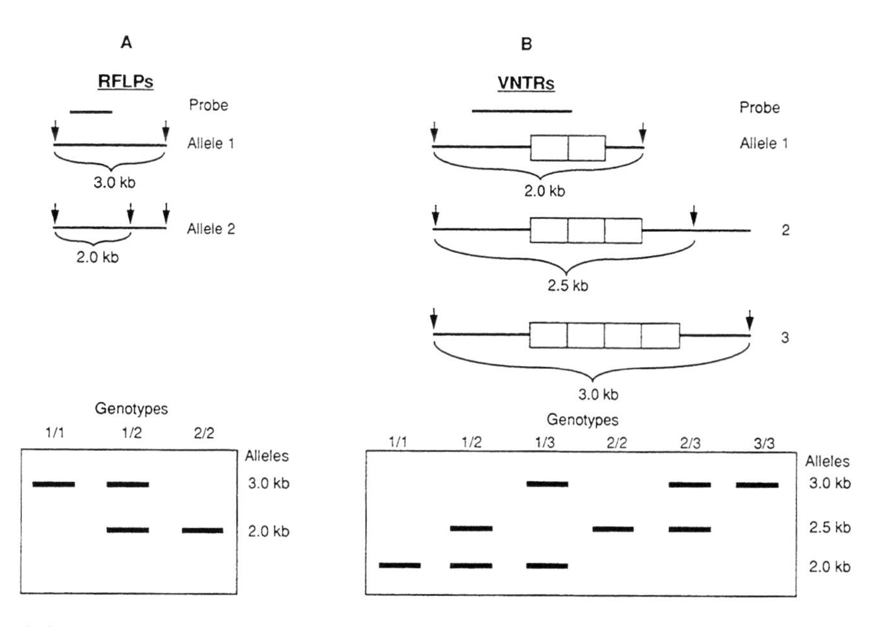

FIGURE 1-3 Two types of RFLPs. Structure of alleles in chromosome is diagrammed at top; arrows indicate sites of cutting by enzyme; lengths of fragments demonstrated by probe (short line above) are given. Electrophoretic patterns are diagrammed below. A. Diallelic RFLP system resulting from single nucleotide change as diagrammed in Figure 1-2. Electrophoretic patterns are those of three genotypes: homozygotes for either allele 1 or allele 2 and 1/2 heterozygote. B. Multiallelic VNTR system. With three alleles as diagrammed, there are six possible genotypes as demonstrated by electrophoresis.

As shown in Figures 1-2 and 1-3A, variation at the cutting site of a restriction enzyme can result in two alternative forms (alleles): the enzyme cuts or it does not. That is called a diallelic system; three genotypes are possible. If a person received the same allele at a particular site (locus) from both parents, the genotype (or person) is said to be homozygous for that allele; if different alleles were inherited from the two parents, the person is heterozygous at that locus. (Some use the term locus, rather than site. Others reserve locus for use in relation to expressed genes.)

As also shown in Figure 1-3B, when the variation is in the number of tandem repeats, there can be many alleles, of which a given person can have only two. That is called a multiallelic, or hypervariable, system. The number of genotypes possible is the sum of the positive integers from 1 to the number of alleles; e.g., in a three-allele system, as shown in Fig. 1-3B, there are expected to be 6 genotypes (1 + 2 + 3). The number of genotypes is also given by the formula $n(n + 1)/2$, where n is the number of alleles.

Inheritance of variation in the noncoding segments of DNA follows the same rules that Mendel inferred for expressed genes. A given individual inherits one of the father's two alleles and one of the mother's two alleles. When two variable sites, each on a different chromosome, are examined, the inheritance at one site is independent of that at the other; i.e., which paternal allele is inherited at site 1 bears no relation to which paternal allele is inherited at site 2. When the two sites are on the same chromosome, they might also be transmitted independently, if they are sufficiently far apart. When they are very close on the same chromosome, the phenomenon of linkage disequilibrium can result—a deviation from independent inheritance in which particular alleles at the two sites tend to be transmitted together.

TECHNOLOGICAL BASIS OF DNA TYPING

Forensic DNA typing usually consists of comparing "evidence DNA" i.e., DNA extracted from material—most often semen—left at a crime scene with "suspect DNA" (i.e., DNA extracted from the blood of a suspect). The tools of DNA typing include restriction enzymes, electrophoresis, probes, and the polymerase chain reaction.[11-13]

Restriction Fragment Length Polymorphisms

In the RFLP approach shown in Figure 1-4, DNA is subjected to controlled fragmentation with restriction enzymes that cut double-stranded DNA at sequence-specific positions. The long DNA molecules are thereby reduced to a reproducible set of short pieces called restriction fragments (RFs), which are usually several hundred to several thousand basepairs long. Many hundreds of thousands of fragments are produced by digestion of human

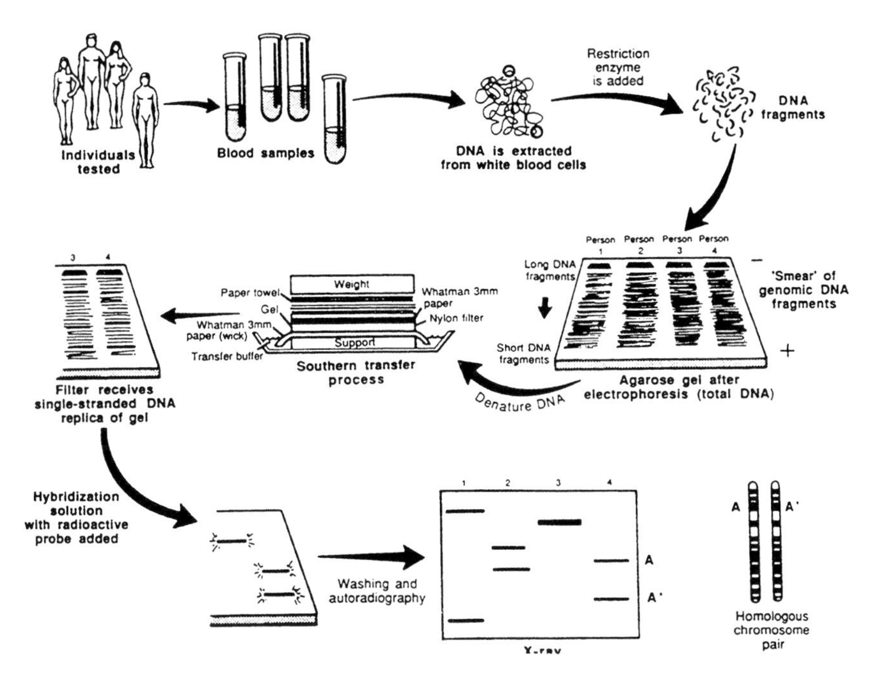

FIGURE 1-4 Schematic representation of Southern blotting of single-locus, multi-allelic VNTR. In example shown here, DNA from four persons is tested. All have different patterns. Three are heterozygous and one homozygous, for a total of seven different alleles. From L. T. Kirby, "DNA Fingerprinting: An Introduction," Stockton Press, New York, 1990. Copyright © 1990 by Stockton Press. Reprinted with permission of W.H. Freeman and Company.

DNA with a single restriction enzyme; each fragment has a distinct sequence and length. For analysis of RFs to demonstrate RFLPs, the fragments are separated electrophoretically on the basis of size. Electrophoresis, typically performed on agarose or acrylamide gels, results in large fragments at one end and small fragments at the other; the small fragments migrate farthest in the electric field. The fragments are then denatured (i.e., rendered single-stranded), neutralized, and transferred from the gel to a nylon membrane, to which they are fixed; this facilitates detection of specific RFLPs and VNTRs.

RFLPs that are defined by specific sequences are detected by hybridization with a probe, a short segment of single-stranded DNA tagged with a group such as radioactive phosphorus, that is used to detect a particular complementary DNA sequence. The nylon membrane is placed in a bath that contains the probe, and the probe hybridizes to the target denatured RF. Nonspecifically bound probe is washed off. Hybridization probes have conventionally been labeled with radioactive isotopes, but attention is increasingly being given to nonisotopic labeling. When isotopically labeled probes are used, the pattern of probe binding is visualized with autoradiography (see examples in Box 1-A and Figure 1-5).

The complete process—DNA digestion, electrophoresis, membrane transfer, and hybridization—was developed by Edwin Southern in 1975;[14] in its present modified form, it is still usually referred to as Southern blotting. These procedures are routinely used in molecular biology, biochemistry, genetics, and clinical DNA diagnosis; there is no difference in their forensic application. Differences among individuals are expressed as differences in the lengths of RFs.[15] RFLPs can result from several kinds of differences at the level of the genome:

- Mutations that alter the base sequence at a restriction-enzyme recognition, or cleavage, site can result in a loss of the cutting site or the generation of a cutting site that was not present before. Insertion or deletion of nucleotides between two cleavage sites also changes RF lengths. Variation of these sorts is generally associated with a small number of alleles. For example, the loss or gain of a particular cleavage site might be responsible for only two alleles.
- Some regions of DNA contain multiple segments of short-sequence repeats. Consequently, there is a class of RFs that differ in the number of repeated segments present. Some VNTR polymorphisms have a small number of alleles, and the patterns of RFs that represent each of the alleles at a given locus can be readily distinguished. But highly polymorphic VNTR loci have 50-100 alleles or even more. In that situation, the distribution of RF size is essentially continuous; alleles with RFs close in size might not be resolvable with electrophoresis, and the limit of resolution must be defined operationally. Because of the extensive variability, the VNTR class of RFLPs has proved the most informative in distinguishing among persons.

RFLP analysis with single-locus probes is usually designed to result in a simple pattern of one or two RFLP bands, depending on whether the

---

FIGURE 1-5 One of several autoradiographs generated during course of investigation of actual rape-murder case. Law enforcement officials submitted blood samples from victim and suspect and vaginal swabs from victim. DNA was isolated from stains found on vaginal swabs, digested with restriction enzyme *Pst*I, and hybridized to probes at genetic loci D2S44, D17S79, D14S13, and D18S27 and monomorphic locus DXZ-1. Known samples are represented in lanes labeled "victim (known)" and "suspect (known)." Lanes labeled "male fraction" and "female fraction" represent DNA resulting from differential lyses of vaginal swabs. "Sensitivity control" lane and "K562 cell line" lane contain 50 ng and 1 μg, respectively, of DNA from immortalized cell line. "Combination marker" lane contains "cocktail" of molecular-weight marker and adenovirus (to monitor migration during electrophoresis).

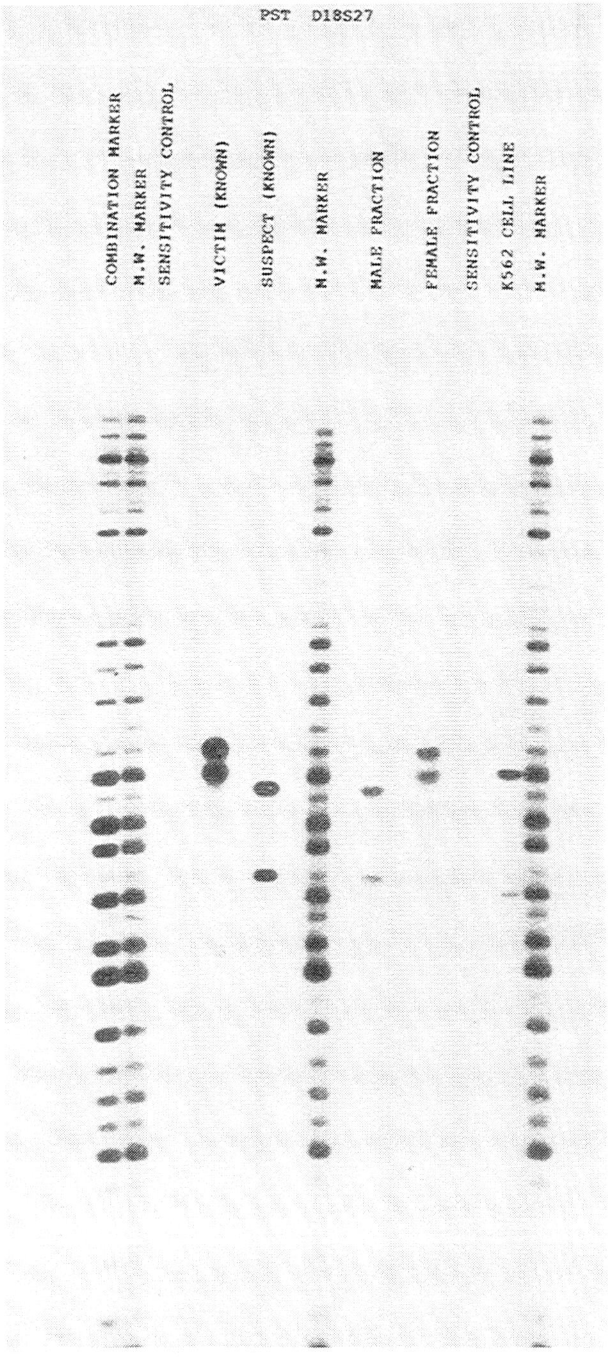

PST D18S27

COMBINATION MARKER
M.W. MARKER
SENSITIVITY CONTROL
VICTIM (KNOWN)
SUSPECT (KNOWN)
M.W. MARKER
MALE FRACTION
FEMALE FRACTION
SENSITIVITY CONTROL
K562 CELL LINE
M.W. MARKER

person is homozygous or heterozygous, respectively. The range of variation shown in the patterns from different persons depends on how many different alleles exist at the particular target locus, e.g., how many different tandem repeats are in the population as a whole (see Figure 1-3B).

An alternative to the use of a single-locus probe is the use of a multilocus probe that hybridizes to many different VNTR sites in the genome. The resulting patterns in a single person contain many bands of varied intensity; the patterns have been compared with barcodes. The approach was developed by Jeffreys and colleagues.[2,3,16,17] Because of the complexity of the patterns, interpretation can be difficult. Consequently, the use of multiple single-locus probes is favored.

Polymerase Chain Reaction for Amplifying DNA

Use of the polymerase chain reaction (PCR) allows a million or more copies of a short region of DNA to be made. It is a method of DNA amplification. For DNA typing, one amplifies a genetically informative sequence, usually 100-2,000 bp long, and detects the genotype in the amplified product. Because many copies are made, genetic typing can rely on nonisotopic methods. With PCR amplification, very small samples of tissue or body fluids—theoretically even a single nucleated cell—can be used to study DNA.[18-19]

The PCR process (Figure 1-6) is simple; indeed, it is analogous to the process by which cells replicate their DNA.[20-23] Two short oligonucleotides are hybridized to the opposite strands of a target DNA segment in positions flanking the sequence region to be amplified; the two oligonucleotides are oriented so that their 3' ends point toward each other. (The ends of a DNA segment are referred to as 5' and 3'; synthesis of new chains proceeds from the 3' end.) The two oligonucleotides serve as primers for an enzyme-mediated replication of the target sequence. The PCR amplification process itself consists of a three-step cycle:

1. The double-stranded template DNA is dissociated into single strands by incubation at high temperature, typically 94°C.
2. The temperature is lowered to allow the oligonucleotide primers to bind to their complementary sequences in the DNA that is to be amplified.
3. A DNA polymerase extends the primers from each of the two primer-binding sites across the region between them, with the target sequence as template.

Because the extension products of one primer bind the other primer in successive cycles, there is in principle a doubling of the target sequence in each cycle. However, the efficiency of amplification is not 100%, and the yield from a 30-cycle amplification is generally about 10^6-10^7 copies of the

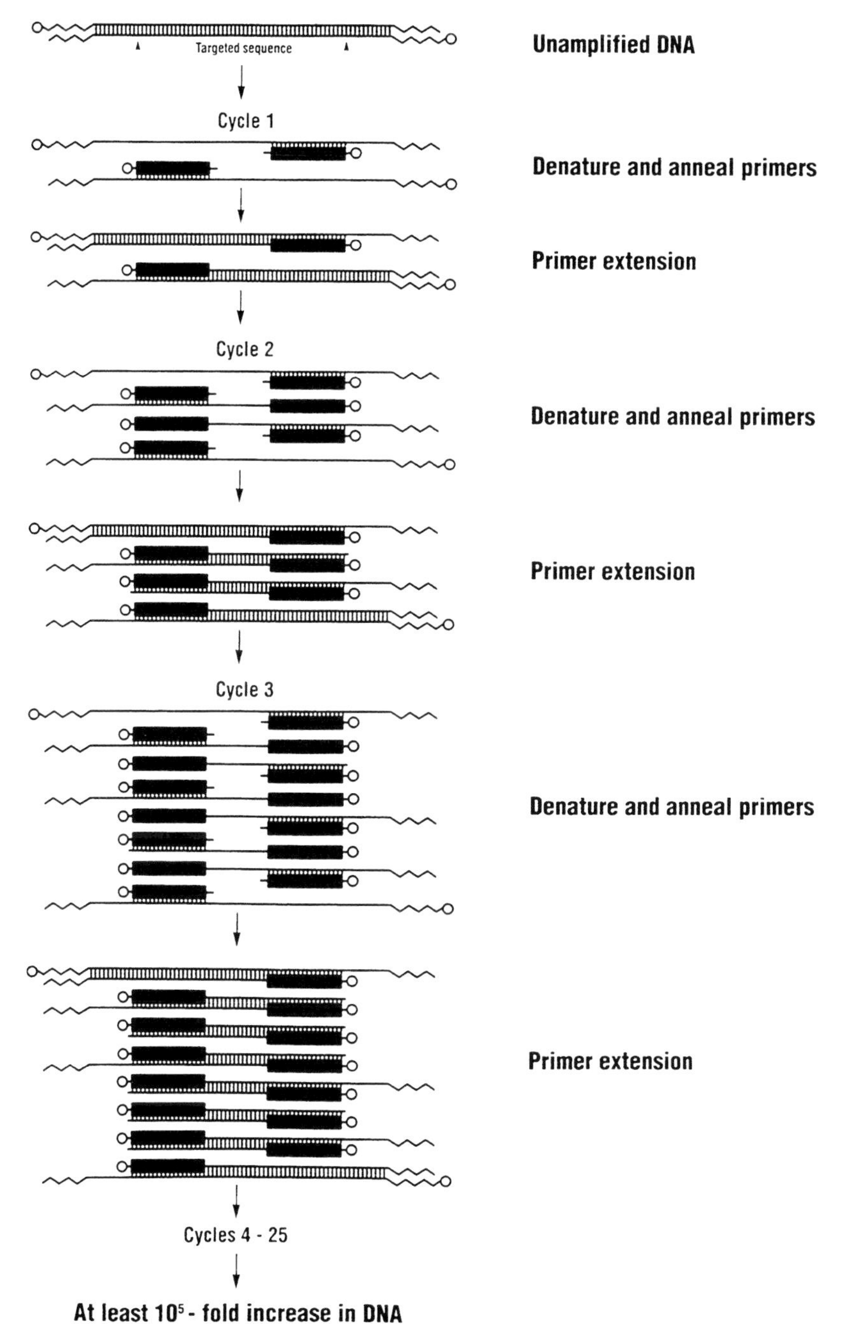

FIGURE 1-6 Polymerase chain reaction (PCR). Courtesy, Perkin-Elmer Cetus Instruments.

target sequence. The primers become physically incorporated into the amplification products.

Both the amplification and the genetic typing can be completed in a day. The efficiency can be improved by amplifying several different products in the same reaction mix; this is termed multiplex amplification.

Several methods have been coupled with PCR for the detection of genetic variation in the amplified DNA; some are listed in Table 1.2. Detection of variation in DNA with PCR-amplified material is fundamentally no different from detection with unamplified samples. The difference is technical: the availability of the larger quantity of pure material produced with PCR affords more options for means of detection.

The use of allele-specific oligonucleotide (ASO) probes is the most generalized approach to the detection of alleles that differ in sequence.[33] The sequence-specific probe is usually a short oligonucleotide, 15-30 nucleotides long, with a sequence exactly matching the sequence of the target allele. The ASO probe is mixed with dissociated strands of PCR reaction product under such conditions that the ASO and the PCR product strands hybridize if there is perfect sequence complementarity, but do not if there are mismatches in sequence.

The usual format for the use of ASO probes is to spot dissociated PCR product strands onto a nitrocellulose or nylon membrane and probe the membrane with labeled ASO. That is analogous to Southern blotting; because the samples are spotted as a "dot" on the membrane, the method is referred to as "dot blotting" (Figure 1-7). An alternative method uses an array of ASO probes immobilized on a test strip.[34] The test strip is immersed in a solution of labeled PCR product; the PCR product hybridizes only to its complementary probe. This procedure has been called "reverse dot blotting" or "blot dotting." A commercial kit based on the reverse dot blot principle has been released (Cetus Corporation).

TABLE 1.2 Some PCR-Based Systems for the Detection of Genetic Variation

Sequence-based detection systems
Allele-specific oligonucleotide (ASO)[24]
Allele-specific priming of PCR[25-27]
Oligonucleotide-ligation assay (OLA)[28]
Restriction-site-specific cleavage (Amp-FLPs)[29]
Denaturing-gradient gel electrophoresis[30]
Chemical cleavage of mismatched heteroduplexes[31]
Length-variation systems
Simple insertions and deletions
VNTR polymorphisms[32]
Analysis of nucleotide sequences

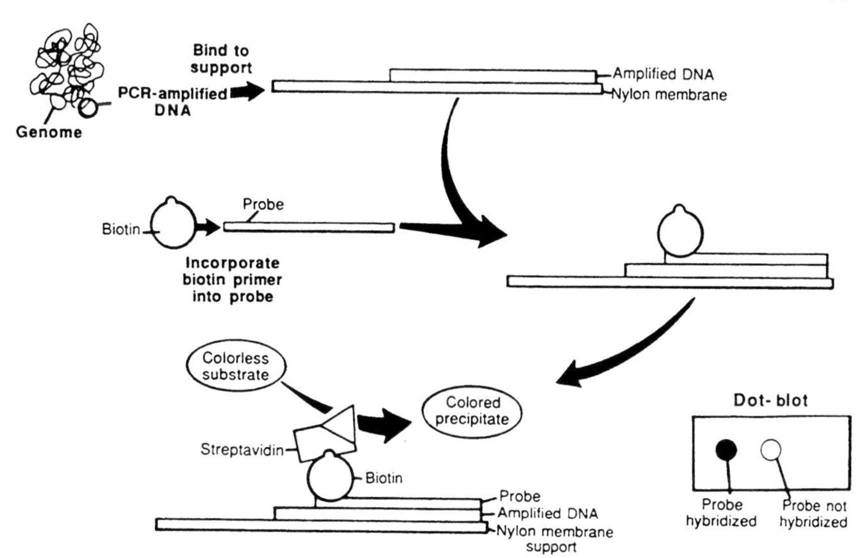

FIGURE 1-7 Diagrammatic representation of dot-blot procedure. PCR-amplified target DNA is immobilized on nylon membrane, and biotinylated probe hybridizes to the target if there is no nucleotide mismatch. Avidin-horseradish peroxidase (HPR) conjugate that binds to biotinylated probe is added. HPR converts colorless substrate to colored product. From L. T. Kirby, "DNA Fingerprinting: An Introduction," Stockton Press, New York, 1990. Copyright © 1990 by Stockton Press. Reprinted with permission of W.H. Freeman and Company.

Amplification of DNA regions that contain inserts or deletions yields products of different size, which are readily detected with electrophoresis. It has proved possible to amplify some of the discrete alleles that make up VNTR polymorphisms;[19] these typing systems combine convenience with the potential for good discrimination. A number of PCR-VNTR typing systems are in development, and it is anticipated that they will come into greater use over the next few years.[35] Indeed, the whole technology of DNA typing can be expected to evolve continually in the next decade. A recent description of PCR-based "digital" DNA typing by Jeffreys is but one example of what might come.[36] In his method, PCR is used to amplify tandem repeats in a portion of DNA. These repeats are then analyzed for the presence of nucleotide polymorphism and assigned numerically (digits 1,2,3, etc.). Thus, by analyzing the repeats (50 or more repeats can easily be analyzed), one is able to assign a specific digital code to each sample (e.g., 1122113111221112...) and unambiguously distinguish one sample from another. The method is simple, avoids match criteria, and requires no side-by-side comparison of DNA samples. The Jeffreys "digital" DNA typing method is still in the research stage, but, if perfected and adapted for forensic samples, it might be possible to achieve absolute identification of persons by analysis of a few such repeats.

PCR provides excellent starting material for direct DNA sequencing, and sequence analysis might ultimately be the approach used for personal identification. But it will require improvements in automated sequencing technology and the generation of larger databases on sequence variability.

POPULATION GENETICS RELEVANT TO THE INTERPRETATION OF DNA TYPING

The finding that two samples of human tissue differ in their DNA patterns leads to the conclusion that the two came from different persons. However, if the two samples are indistinguishable with regard to the detected DNA patterns, two possibilities exist: the two samples came from the same person (or from identical twins), or the two samples came from different persons whose DNA patterns in the target regions investigated are the same. Those two possibilities cannot be distinguished. To provide the trier of fact—a judge or jury, for example—with information for weighing the two possibilities, it has been traditional to use statistics from population genetics to estimate the fraction of people in the population who have the particular combination of DNA patterns. What is the chance of picking at random a person who has the same genetic patterns as found in the evidence sample? Obviously, the lower the probability, the stronger the inference that the evidence sample is associated with a particular person who has those patterns.

Thus, the central question in relation to forensic DNA typing is, What is the probability that a person picked at random would match the evidence sample in DNA patterns? Or, What proportion of persons in the same population as a suspect have the same combination of DNA patterns as the evidence sample? In answering the questions, the population frequencies of the patterns at the several (often four) loci tested are multiplied together, on the assumption that they are independent. Indeed, the alleles (two in a diallelic system) at each locus are assumed to vary independently. Estimation of the frequencies of specific alleles needed to answer the questions is based on population genetics.[37] Experience with blood groups, enzyme markers, and HLA types—including the genetic markers used for forensic personal "identification" in the pre-DNA era—indicates that the frequency of specific allele can vary widely among populations. That is demonstrated in Figure 1-8 for the B blood-group allele; maps of the A and O blood-group alleles show similar variability. It is to be expected that the frequencies of the various alleles for the DNA polymorphisms also show differences among populations. Thus, it might be critically important to pick the right population with which to compare a given suspect's DNA.

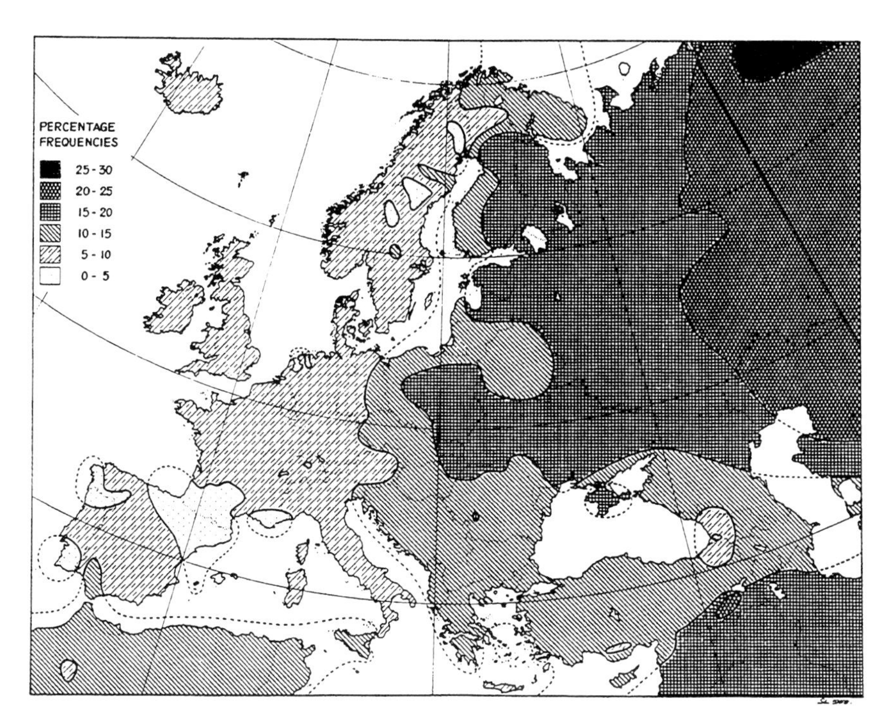

FIGURE 1-8 Distribution of gene for blood group B in Europe. Similar wide variation is observed with genes for blood groups A and O. From Mourant et al.,[38] p. 266.

Estimating the Frequency of Alleles in Populations

As diagrammed in Figure 1-9, a diallelic system such as represented by many RFLPs (Figure 1-3) gives rise to three genotypes that in the state of Hardy-Weinberg equilibrium (see Chapter 3) are expected to have the following frequencies: p^2 (homozygotes for allele A1), 2pq (heterozygotes), and q^2 (homozygotes for allele A2), where p is the frequency of allele A1 and q is the frequency of allele A2. The frequency of each of the two alleles can be derived by counting: e.g., each A1 homozygote has 2 A1 alleles, and each A1/A2 heterozygote has 1 A1 allele. Or the proportion of the A1 allele can be taken as the square root of the frequency of A1 homozygotes. The two methods should give closely similar results, if the population is in Hardy-Weinberg equilibrium.

A multiallelic system like that represented by VNTRs (Figure 1-10) gives rise to many genotypes, as diagrammed in Figure 1-11; with five alleles, there are 15 genotypes [n(n + 1)/2]. Again, the frequency of each allele can be determined by counting or, more easily, by taking the square

Allele from mother \ Allele from father	A1(p)	A2(q)
A1 (p)	A1/A1 (p^2)	A1/A2 (pq)
A2 (q)	A1/A2 (pq)	A2/A2 (q^2)

p,q = proportion of A1 and A2 alleles, respectively, in the population -- p + q = 1

p^2, 2pq, q^2 = proportion of 3 genotypes -- $p^2 + 2pq + q^2 = 1$

where p^2 = proportion of A1 homozygotes

2pq = proportion of heterozygotes

q^2 = proportion of A2 homozygotes

FIGURE 1-9 Punnett square (named for British geneticist) indicating frequency of genotypes in diallelic RFLP system.

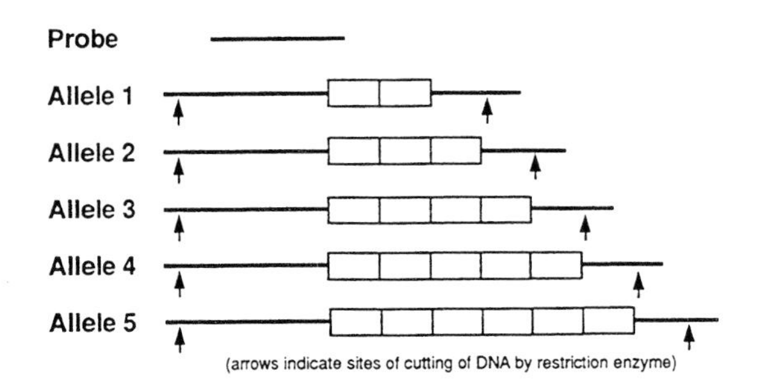

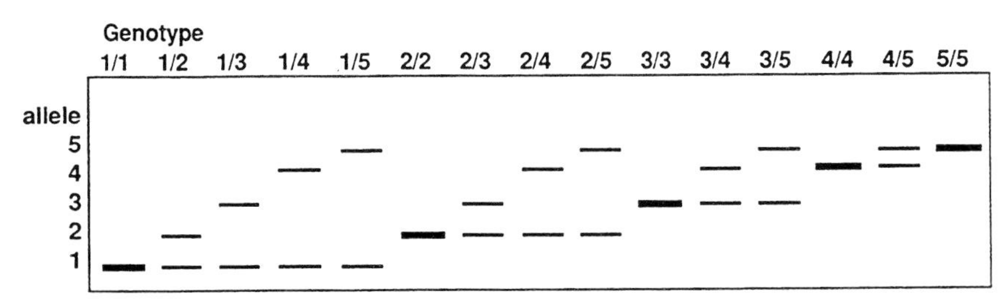

FIGURE 1-10 Schematic representation of alleles and genotypes in five-allele VNTR system.

A

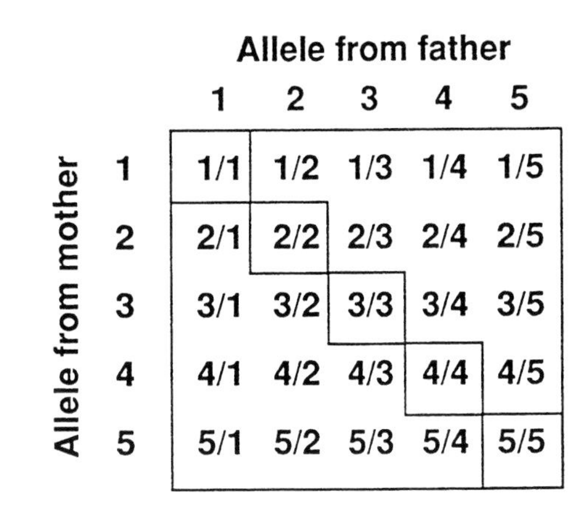

p = frequency of allele 1
q = frequency of allele 2
r = frequency of allele 3
s = frequency of allele 4
t = frequency of allele 5

15 genotypes from 5 alleles (5 + 4 + 3 + 2 + 1, or $n(n+1)/2$)

Frequency of genotypes

5 homozygotes	10 heterozygotes			
1/1 = p^2	1/2 = 2pq	2/3 = 2qr	3/4 = 2rs	4/5 = 2st
2/2 = q^2	1/3 = 2pr	2/4 = 2qs	3/5 = 2rt	
3/3 = r^2	1/4 = 2ps	2/5 = 2qt		
4/4 = s^2	1/5 = 2pt			
5/5 = t^2				

B

Worked Example

Given:

p = 30%
q = 30%
r = 20%
s = 10%
t = 10%

Frequency of genotypes

Homozygotes	Heterozygotes			
p^2=9%	2pq=18%	2qr=12%	2rs=4%	2st=2%
q^2=9%	2pr=12%	2qs=6%	2rt=4%	
r^2=4%	2ps=6%	2qt=6%		
s^2=1%	2pt=6%			
t^2=1%				

FIGURE 1-11 A. Punnett square showing genotypes in single locus, five-allele VNTR system, as schematized in Figure 1-10. B. A worked example: frequency of each of 15 genotypes with allele frequencies indicated.

root of the frequency of the homozygotes for the particular allele or half the summed frequencies of the heterozygotes, which, in the case of the five-allele system, will be of four types for each allele.

Population Substructure

It is intuitively obvious that relatives have genes in common. Thus, the chance that DNA typing will yield a match with a suspect when the evidence sample in fact came from a brother or other relative is considerable, especially if only a few loci are tested.

The U.S. population is a conglomerate of many different population groups, which might be viewed as extended families derived from all parts of the world. The major ethnic groupings—white, black, Hispanic, Asian, etc.—are each composites of many different subpopulations, which might have quite different frequencies of the alleles used in forensic DNA typing. Allele frequencies estimated from sampling of an overall ethnic group represent weighted averages. Some of the component subpopulations might have allele frequencies quite different from the mean values of the whole population. Further discussion of this problem and recommendations for handling it are given in Chapter 3.

CHARACTERISTICS OF AN OPTIMAL FORENSIC DNA TYPING SYSTEM

The methods of DNA typing continue to evolve as new ways to detect individual variation are developed. Sequencing of DNA might ultimately be the optimal method of personal identification, but that is still far from practical. It is important that the flexibility to adopt new methods be retained as standardization of DNA technology is developed (see Chapter 4) and databanks are created (see Chapter 5).

Any method of forensic DNA typing, like methods for medical DNA and other testing, should be rapid, accurate, and inexpensive. In addition, to achieve maximal discrimination among individuals, forensic DNA typing requires the use of markers with a high level of variability or polymorphism. Ideally, the high degree of variability would be found in all the world's populations. The markers and the probes used to detect them should have a unique sequence, so that each probe hybridizes with only one part of the genome. Single-locus probes should be used. The loci of the markers should be independent, e.g., on separate chromosomes. The markers should, furthermore, come from noncoding and therefore presumably nonfunctional parts of the genome, to avoid claims, spurious or otherwise, of association of particular markers with particular behavioral traits or diseases.

The automation of DNA typing might help to reduce its time and expense. An advantage of speed and low cost is that one can test more parts of the genome. Even if a locus is only modestly polymorphic, its use in DNA typing could have other advantages, such as complete unambiguity of scoring; used in combination, such loci could demonstrate that the chance of a random match is extremely low.

It must be emphasized that new methods and technology for demonstrating individuality in each person's DNA continue to be developed. The methods outlined in this chapter are likely to be superseded in efficiency, automatability, economy, and other features by new methods. Care must be taken to ensure that DNA typing techniques used for forensic purposes do not become "locked in" prematurely. Otherwise, society and the criminal justice system will not be able to derive maximal benefit from advances in the science and technology.

REFERENCES

1. Gaensslen RE. Sourcebook in forensic serology, immunology, and biochemistry. Washington, D.C.: U.S. Government Printing Office, 1983.
2. Jeffreys AJ, Wilson V, Thein SL. Individual-specific "fingerprints" of human DNA. Nature. 316:75-79, 1985.
3. Gill P, Jeffreys AJ, Werrett DJ. Forensic application of DNA "fingerprints." Nature. 318:577-579, 1985.
4. Evett IW, Werrett DJ, Gill P, Buckleton JS. DNA fingerprinting on trial. Nature. 340:435, 1989.
5. Lander ES. DNA fingerprinting on trial. Nature. 339:501-505, 1989.
6. U.S. Congress, Office of Technology Assessment. Genetic witness: forensic uses of DNA tests. OTA-BA-438. Washington, D.C.: U.S. Government Printing Office, 1990.
7. Thompson L. A smudge on DNA fingerprinting? The Washington Post, Monday, June 26, 1989.
8. Barinaga M. DNA fingerprinting: pitfalls come to light. Nature. 339:89, 1989.
9. Roychoudhury AK, Nei M. Human polymorphism genes: world distribution. New York: Oxford University Press, 1988.
10. Cooper DN, Smith BA, Cooke HJ, Niemann S, Schmidtke J. An estimate of unique DNA sequence heterozygosity in the human genome. Hum Genet. 69:201-205, 1985.
11. Kirby LT. DNA fingerprinting: an introduction. New York: Stockton Press, 1990.
12. Farley MA, Harrington JJ, eds.: Forensic DNA technology. Chelsea, Michigan: Lewis Publishers, 1991.
13. U.S. Department of Justice, Federal Bureau of Investigation. Proceedings of the international symposium on the forensic aspects of DNA analysis. Washington, D.C.: U.S. Government Printing Office, 1991.
14. Southern EM. Detection of specific sequences among DNA fragments separated by gel electrophoresis. J Mol Biol. 98:503-527, 1975.
15. Botstein D, White RL, Skolnick M, Davis RW. Construction of a genetic linkage map in man using restriction fragment length polymorphisms. Am J Hum Genet. 32:314-331, 1980.
16. Helminen P, Ehnholm C, Lokki ML, Jeffreys A, Peltonen L. Application of DNA "fingerprints" to paternity determinations. Lancet. 1:574-576, 1988.
17. Jeffreys AJ, Wilson V, Thein SL, Weatherall DJ, Ponder BAJ. DNA "fingerprints" and segregation analysis of multiple markers in human pedigrees. Am J Hum Genet. 39:11-24, 1986.

18. Higuchi R, von Beroldingen CH, Sensabaugh GF, Erlich HA. DNA typing from single hairs. Nature. 332:543-546, 1988.
19. Jeffreys AJ, Wilson V, Neumann R, Keyte J. Amplification of human minisatellites by the polymerase chain reaction: towards DNA fingerprinting of single cells. Nucleic Acids Res. 16:10953-10971, 1988.
20. Erlich HA, ed. PCR technology: principles and applications for DNA amplification. New York: Stockton Press, 1989.
21. Rose EA. Applications of the polymerase chain reaction to genome analysis. FASEB J. 5:46-54, 1991.
22. Arnheim N, Levenson CH. Polymerase chain reaction. Chem Eng News. 68:36-47, October 1, 1990.
23. Mullis KB. The unusual origin of the polymerase chain reaction. Sci Am. 262:56-65, 1990.
24. Conner BJ, Reyes AA, Morin C, Itakura K, Teplitz RL, Wallace RB. Detection of sickle-cell beta-S-globin allele by hybridization with synthetic oligonucleotides. Proc Natl Acad Sci USA. 80:278-282, 1983.
25. Chehab FF, Kan YW. Detection of specific DNA-sequences by fluorescence amplification—a color complementation assay. Proc Natl Acad Sci USA. 86:9178-9182, 1989.
26. Newton CR, Graham A, Heptinstall LC, Porvell SJ, Summers JC, Markham AF. Analysis of any point mutation in DNA—the amplification refractory mutation system (ARMS). Nucleic Acids Res. 17:2503-2516, 1989.
27. Wu DY, Ugozzoli L, Pal BK, Wallace RB. Allele-specific enzymatic amplification of beta-globin genomic DNA for diagnosis of sickle-cell anemia. Proc Natl Acad Sci USA. 86:2757-2760, 1989.
28. Nickerson DA, Kaiser R, Lappin S, Steward J, Hood L, Landegren U. Automated DNA diagnostics using an ELISA-based oligonucleotide ligation assay. Proc Natl Acad Sci USA. 87:8923-8927, 1990.
29. Kasai K, Nakamura Y, White R. Amplification of a variable number of tandem repeats (VNTR) locus (pMCT118) by the polymerase chain reaction (PCR) and its application to forensic science. J For Sci. 35:1196-1200, 1990.
30. Myers RM, Maniatis T, Lerman LS. Detection and localization of single base changes by denaturing gradient gel electrophoresis. Methods Enzymol. 155:501-527, 1987.
31. Cotton RGH, Rodriques NR, Campbell RD. Reactivity of cytosine and thymine in single-base-pair mismatches with hydroxylamine and osmium-tetroxide and its application to the study of mutations. Proc Natl Acad Sci USA. 85:4397-4401, 1988.
32. Nakamura Y, Leppert M, O'Connell P, Wolff R, Holm T, Culver M, Martin C, Fujimoto E, Hoff M, Kumlin E, White R. Variable number of tandem repeat (VNTR) markers for human gene mapping. Science. 235:1616-1622, 1987.
33. Saiki R, Bugawan TL, Horn GT, Mullis KB, Erlich HA. Analysis of enzymatically amplified β-globin and HLA-DQ DNA with allele-specific oligonucleotide probes. Nature. 324:163-166, 1986.
34. Saiki R, Walsh PS, Levenson CH, Erlich HA. Genetic analysis of amplified DNA with immobilized sequence-specific oligonucleotide probes. Proc Natl Acad Sci USA. 86:6230-6234, 1989.
35. Hagelberg E, Gray IC, Jeffreys AJ. Identification of the skeletal remains of a murder by DNA analysis. Nature. 352:427-428, 1991.
36. Jeffreys A, MacLeod A, Tamaki K, Neil D, Monckton D. Minisatellite repeat coding as a digital approach to DNA typing. Nature. 354:204-209, 1991.
37. Hartl DL, Clark AC. Principles of population genetics. 2nd ed. Sunderland, Massachusetts: Sinauer Associates, 1989.
38. Mourant AE, Kopec AC, Domaniewska-Sobczak K. The distribution of the human blood groups and other polymorphisms. 2nd ed. Oxford: Oxford University Press, 1976.

2

DNA Typing: Technical Considerations

"DNA typing" is a catch-all term for a wide range of methods for studying genetic variations. Each method has its own advantages and limitations, and each is at a different state of technical development. Each DNA typing method involves three steps:

1. Laboratory analysis of samples to determine their genetic-marker types at multiple sites of potential variation.
2. Comparison of the genetic-marker types of the samples to determine whether the types match and thus whether the samples could have come from the same source.
3. If the types match, statistical analysis of the population frequency of the types to determine the probability that such a match might have been observed by chance in a comparison of samples from different persons.

Before any particular DNA typing method is used for forensic purposes, it is essential that precise and scientifically reliable procedures be established for performing all three steps. This chapter discusses the first two—laboratory analysis and pattern comparison—and Chapter 3 focuses on statistical analysis.

There is no scientific dispute about the validity of the general principles underlying DNA typing: scientists agree that DNA varies substantially among humans, that variation can be detected in the laboratory, and that DNA comparison can provide a basis for distinguishing samples from different persons. However, a given DNA typing method might or might not be scientifically appropriate for forensic use. Before a method can be accepted as valid for forensic use, it must be rigorously characterized in both research and forensic settings to determine the circumstances under which it will and will not yield reliable results. It is meaningless to speak of the reliability of DNA typing in general—i.e., without specifying a particular method. Some states have adopted vaguely worded statutes regarding admissibility of DNA typing results without specifying the methods intended to be covered. Such laws obviously were intended to cover only conventional RFLP analysis of single-locus probes on Southern blots—the only method in common use at the time of passage of the legislation. We trust that courts will recognize the limitations inherent in such statutes.

Forensic DNA analysis should be governed by the highest standards of scientific rigor in analysis and interpretation. Such high standards are appropriate for two reasons: the probative power of DNA typing can be so great that it can outweigh all other evidence in a trial; and the procedures for DNA typing are complex, and judges and juries cannot properly weigh and evaluate conclusions based on differing standards of rigor.

The committee cannot provide comprehensive technical descriptions for DNA typing in this report: too many methods exist or are planned, and too many issues must be addressed in detail for each method. Instead, our main goal is to provide a general framework for the evaluation of any DNA typing method.

ESSENTIALS OF A FORENSIC DNA TYPING PROCEDURE

Scientific Foundations

The forensic use of DNA typing is an outgrowth of its medical diagnostic use—analysis of disease-causing genes based on comparison of a patient's DNA with that of family members to study inheritance patterns of genes or with reference standards to detect mutations. To understand the challenges involved in such technology transfer, it is instructive to compare forensic DNA typing with DNA diagnostics.

DNA diagnostics usually involves clean tissue samples from known sources. It can usually be repeated to resolve ambiguities. It involves comparison of discrete alternatives (e.g., which of two alleles did a child inherit from a parent?) and thus includes built-in consistency checks against artifacts. It requires no knowledge of the distribution of patterns in the general population.

Forensic DNA typing often involves samples that are degraded, contaminated, or from multiple unknown sources. It sometimes cannot be repeated, because there is too little sample. It often involves matching of samples from a wide range of alternatives present in the population and thus lacks built-in consistency checks. Except in cases where the DNA evidence excludes a suspect, assessing the significance of a result requires statistical analysis of population frequencies.

Despite the challenges of forensic DNA typing, we believe that it is possible to develop reliable forensic DNA typing systems, provided that adequate scientific care is taken to define and characterize the methods. We outline below the principal issues that must be addressed for each DNA typing procedure.

Written Laboratory Protocol

An essential element of any clinical or forensic DNA typing method is a detailed written laboratory protocol. Such a protocol should not only specify steps and reagents, but also provide precise instructions for interpreting results, which is crucial for evaluating the reliability of a method. Moreover, the complete protocol should be made freely available so that it can be subjected to scientific scrutiny.

Procedure For Identifying Patterns

There must be an objective and quantitative procedure for identifying the pattern of a sample. Although the popular press sometimes likens DNA patterns to bar codes, laboratory results from most methods of DNA testing are not discrete data, but rather continuous data. Typically, such results consist of an image—such as an autoradiogram, a photograph, spots on a strip, or the fluorometric tracings of a DNA sequence—and the image must be quantitatively analyzed to determine the genotype or genotypes represented in the sample. Quantitation is especially important in forensic applications, because of the ever-present possibility of mixed samples.

Patterns must be identified separately and independently in suspect and evidence samples. It is not permissible to decide which features of an evidence sample to count and which to discount on the basis of a comparison with a suspect sample, because this can bias one's interpretation.

Procedure For Declaring a Match

When individual patterns of DNA in evidence sample and suspect sample have been identified, it is time to make comparisons to determine whether they match. Whether this step is easy or difficult depends on the resolving power of the system to distinguish alleles. Some DNA typing methods involve small collections of alleles that can be perfectly distinguished from one another—e.g., a two-allele RFLP system based on a polymorphism at a single locus. Other methods involve large collections of similar alleles that are imperfectly distinguished from one another—e.g., the hypervariable VNTR systems in common forensic use, in which a single sample might yield somewhat different allele sizes on repeat measurements.[1] It is easy to determine whether two samples match in the former case (assuming that the patterns have been correctly identified), but the latter case requires a match criterion—i.e., an objective and quantitative rule for deciding whether two samples match. For example, a match criterion for VNTR systems might declare a match between two samples if the restriction-fragment sizes lie within 3% of one another.

The match criterion must be based on the actual variability in measurement observed in appropriate test experiments conducted in each testing laboratory. The criterion must be objective, precise, and uniformly applied. If two samples lie outside the matching rule, they must be declared to be either "inconclusive" or a "nonmatch." Considerable controversy arose in early cases over the use of subjective matching rules (e.g., comparison by eye) and the failure to adhere to a stated matching rule.

Identification of Potential Artifacts

All laboratory procedures are subject to potential artifacts, which can lead to incorrect interpretation if not recognized. Accordingly, each DNA typing method must be rigorously characterized with respect to the types of possible artifacts, the conditions under which they are likely to occur, the scientific controls for detecting their occurrence, and the steps to be taken when they occur, which can range from reinterpreting results to correcting for the presence of artifacts, repeating some portion of the experiment, or deciding that samples can be reliably used.

Regardless of the particular DNA typing method, artifacts can alter a pattern in three ways: Pattern A can be transformed into Pattern B, Pattern A can be transformed into Pattern A + B; and Pattern A + B can be transformed into Pattern B. It is important to identify the circumstances under which each transformation can occur, because only then can controls and corrections be devised. For example, RFLP analysis is subject to such artifacts as band shifting, in which DNA samples migrate at different speeds and yield shifted patterns (A→B), and incomplete digestion, in which the failure of a restriction enzyme to cleave at all restriction sites results in additional bands (A→A + B).

Some potential problems can be identified on the basis of the chemistry of DNA and the mechanism of detection in the genetic-typing system. Anticipation of potential sources of DNA typing error allows systematic empirical investigation to determine whether a problem exists in practice. If so, the range of conditions in which an assay is subject to artifact must be characterized. In either case, the results of testing for artifacts should be documented. Empirical testing is necessary, whether one is considering a new method, a new locus, a new set of reagents (probe or enzyme) for a pre-existing locus, or a new device. Under some circumstances, even small changes in procedure can change the pattern of artifacts.

Once potential artifacts have been identified, it is necessary to design scientific controls to serve as internal checks in each experiment to test whether the artifacts have occurred. Once the appropriate controls are identified, analysts must use them consistently when interpreting test results. If the appropriate control has not been performed, no result should be reported. When a control indicates irregularities in an experiment, the results in question must be considered inconclusive; if possible, the experiment should be repeated. A well-designed DNA typing test should be a matter of standardized, objective analysis.

Sensitivity to Quantity, Mixture, and Contamination

Evidence samples might contain very little DNA, might contain a mixture of DNA from multiple sources, and might be contaminated with chemicals that can interfere with analysis. It is essential to understand the limits of each DNA typing method under such circumstances.

Experiential Foundation

Before a new DNA typing method can be used, it requires not only a solid scientific foundation, but also a solid base of experience in forensic application. Traditionally, forensic scientists have applied five steps to the implementation of genetic marker systems:[2,3]

1. Gain familiarity with a system by using fresh samples.
2. Test marker survival in dried stains (e.g., bloodstains).
3. Test the system on simulated evidence samples that have been exposed to a variety of environmental conditions.
4. Establish basic competence in using the system through blind trials.
5. Test the system on nonprobative evidence samples whose origin is known, as a check on reliability.

When a technique is initially developed, all five steps should be carefully followed. As laboratories adopt the technique, it will not always be necessary for them to repeat all the steps, but they must demonstrate familiarity and competence by following steps 1, 4, and 5.[4]

Most important, there is no substitute for rigorous external proficiency testing via blind trials. Such proficiency testing constitutes scientific confirmation that a laboratory's implementation of a method is valid not only in theory, but also in practice. No laboratory should let its results with a new DNA typing method be used in court, unless it has undergone such proficiency testing via blind trials. (See Chapter 4 for discussion of proficiency testing.)

Publication and Scientific Scrutiny

If a new DNA typing method (or a substantial variation on an existing one) is to be used in court, publication and scientific scrutiny are very important. Extensive empirical characterization must be undertaken. Results must be published in appropriate scientific journals. Publication is the mechanism that initiates the process of scientific confirmation and eventual acceptance or rejection of a method.

Some of the controversy concerning the forensic use of DNA typing can be traced to the failure to publish a detailed explanation and justification of methods. Without the benefit of open scientific scrutiny, some testing laboratories initially used methods (for such fundamental steps as identifying patterns, declaring matches, making comparison with a databank, and correcting for band shifting) that they later agreed were not experimentally supported. In some cases, those errors resulted in exclusion of DNA evidence or dismissal of charges.

TECHNICAL ISSUES IN RFLP ANALYSIS

Choice of Probes

A DNA probe used in forensic applications should have the following properties:

- It should recognize a single human locus (or site), preferably one whose chromosomal location has been determined.
- It should detect a constant number of bands per allele in most humans.
- It should be characterized in the published literature, including its typical range of alleles, and its tendency to recognize DNA from other species.
- It should be readily available for scientific study by any interested person.

The committee recommends against forensic use of multilocus probes, which detect many fragments per person. Because such probes might detect fragments with quite different intensities, it is difficult to know whether one has detected all fragments in a sample—particularly with small and degraded forensic samples—and difficult to recognize artifacts and mixtures. Such problems increase the difficulty of pattern interpretation. Multilocus probes increase the risk of incorrect interpretation, and numerous single-locus probes, which do not pose such problems, are available. The use of enough single-locus probes gains the advantages of the single multilocus probes without the problems of interpretation.

Southern Blot Preparation

The basic protocol for preparing Southern blots is fairly standard, but testing laboratories vary in such matters as choice of restriction enzyme, gel length and composition, and electrophoresis conditions. Such differences do not fundamentally affect the reliability of the general method, but some enzymes might require characterization (e.g., each restriction enzyme must be characterized for sensitivity to inhibitors, for tendency to cut at anomalous recognition sites under some conditions—often called "star activity"—and for tendency to produce partial digestions), and differences in gels and electrophoresis conditions will affect resolution of fragments and retention of small fragments.

Questions have arisen concerning the use of ethidium bromide, a fluorescent dye that binds to DNA and so allows it to be visualized. Some laboratories incorporate ethidium bromide into analytical gels before electrophoresis; others stain gels with ethidium bromide after electrophoresis. The committee strongly recommends the latter, for two reasons:

• Ethidium bromide binds to DNA in a concentration-dependent manner and has been shown to alter the mobility of fragments at high DNA concentrations, thus decreasing the reliability of fragment-size measurements.

• Staining after electrophoresis requires smaller amounts of ethidium bromide, and that is preferable, because the dye is a known carcinogen and thus poses problems of exposure and disposal.

Because there are several advantages and no drawbacks to staining after electrophoresis, we conclude that there is no present justification for use of ethidium bromide in analytical gels.

Identification of DNA Patterns

Identification of the DNA pattern of each sample should be carried out very carefully. When analyzed with a single-locus probe, each lane will ideally show at the most fragments derived from two alleles and nothing else. However, complications can arise. To interpret such complications properly, an examiner requires considerable knowledge and skill and might need to examine control experiments.

Examination of a Control Pattern

Every Southern blot procedure should be applied to a known DNA sample (in addition to the evidence samples in question), to verify that the hybridization was performed correctly. If this control sample does not yield a clean result that shows the correct pattern for a particular hybridization, the result of the test hybridization should be discounted.

Single-Band Patterns

Sometimes, only a single band will be detected when two distinct alleles are present. That might occur because the second allele is so small that it has migrated off the end of the gel, because the second allele is similar in size to the first allele and thus is not resolved, or because the second allele is much larger, and larger fragments are preferentially lost in partially degraded samples.

When only a single band is found, the interpretation should always include the possibility that a second band has been missed—i.e., that the pattern is actually of a heterozygote, not a homozygote. (For statistical interpretation, the frequency of a single-band pattern should be taken to be the sum of the frequencies of all patterns containing this band. This is approximately twice the allele frequency of the band.) In some cases, it could be important to interpret the absence of a second larger fragment—e.g., when two samples match in a smaller band, but the questioned sample lacks a second larger band. That could arise either because the samples are from different persons or because the samples come from the same person but the questioned sample is partially degraded. Ideally, to distinguish these alternatives, one should determine whether a second larger band *could* have been detected in the questioned sample by hybridizing the membrane with a single-copy probe that detects an even larger monomorphic fragment—i.e., one that is constant in all humans. In contrast, it would not be sufficient simply to estimate the degree of degradation from the ethidium bromide staining pattern of the sample.

Anomalous Bands

A sample might show more than two bands for various reasons. E.g., the hybridization conditions were improper and caused the probe to hybridize to incorrect fragments; the probe was contaminated with another sequence, which caused it to recognize other fragments; the membrane was incompletely stripped after a previous use, so a pattern seen on the previous hybridization is still being detected; the restriction digestion did not proceed to completion, so the region recognized by the probe is present in incompletely cut fragments of multiple sizes; or the sample actually contains a mixture of multiple DNAs. The last example is extremely important to recognize, because it can bear importantly on a case. Whenever extra bands are observed, their origin should be determined.

The following clues provide a partial decision tree:

If the hybridization conditions were improper or the probe contaminated, the pattern in the control DNA should be seen to be incorrect; the hybridization should be repeated.

If the membrane was improperly stripped, the extra bands will be in the same location as in the previous hybridization and might be present in the control sample; the hybridization should be repeated.

If the restriction digestion was incomplete, one should see additional bands, even with the use of monomorphic probes that typically give only a single constant band. To ascribe extra bands to incomplete digestion, one should therefore perform such a hybridization. If incomplete digestion has occurred, the sample ideally should be re-extracted and redigested, and a new Southern blot should be prepared. If that is not possible, because there is too little sample, it will usually be difficult to get a reliable result.

If the samples are mixtures from more than one person, one should see additional bands for all or most polymorphic probes, but not for a single-copy monomorphic probe. Mixed samples can be very difficult to interpret, because the components can be present in different quantities and states of degradation. It is important to examine the results of multiple RFLPs, as a consistency check. Typically, it will be impossible to distinguish the individual genotypes of each contributor. If a suspect's pattern is found within the mixed pattern, the appropriate frequency to assign such a "match" is the sum of the frequencies of all genotypes that are contained within (i.e., that are a subset of) the mixed pattern.

Another possible cause of extra bands is leakage between adjacent sample lanes or misloading of two samples in a single lane. Such an occurrence can be exceedingly difficult to detect and could result in an incorrect conclusion. It is therefore important to leave a blank lane between a suspect sample and an evidence sample, so that leakage can be detected and will not lead to false-positive results.

Reporting of Anomalies

Examiners should document their interpretations of samples thoroughly in writing. They should note all observed bands and any questionable densities that they do not consider to be bands. Anomalous bands should be explained on the basis of appropriate control experiments of the sorts described above.

Measurement of Fragments

Molecular-weight measurements of fragments should initially be made by comparing band positions with known molecular-weight standards run in separate lanes on the same gel (so-called external molecular-weight standards). Measurements should be performed with a computer-assisted or computer-automated system, in which the operator identifies the positions of the bands with a digitizing pen or similar device that directly records them, visually inspects them, or both. Computer-based procedures ensure appropriate documentation of the measurement and promote objectivity.

External molecular-weight standards alone, however, are not sufficient, because anomalies in electrophoresis can lead to errors in RFLP typing caused by band shifting.[5] Such anomalies can be due to differences in salt or DNA concentrations among samples (which could be corrected by repeated extraction) or to covalent or noncovalent modifications of the DNA (which might be irreversible). Band shifting could cause two DNA samples from one person to show different patterns or DNA samples from two different persons to show the same pattern. Band shifting also makes it impossible to measure fragment sizes relative to external molecular-weight standards, because the standards have migrated at a different speed.

Band shifting is easy to detect by hybridizing the Southern blot with monomorphic probes—that is, probes that detect constant-length fragments that are always in the same position in all people. If several monomorphic fragments are in the same position in both lanes, it is safe to assume that no band shifting has occurred. If the monomorphic fragments are in different positions, band shifting is present. The committee considers it desirable for all samples to be tested for band shifting by hybridization with monomorphic probes that cover a wide range of fragment sizes in the gel. That approach will eliminate the rare production of a match by shifting of bands in an evidence sample to the same positions as in a suspect sample. Testing laboratories now investigate the possibility of band shifting only when they find two samples with patterns that appear to be similar but shifted relative to one another. (Multiple monomorphic probes might not be available for some systems and might need to be developed.)

Testing for band shifting is easy, but correcting it is harder. The best approach is to clean the samples (by re-extraction, dialysis, or other measures) and repeat the experiment in the hope of avoiding band shifting. When that is impossible because too little sample is available or it fails (perhaps because of covalent modification of the DNA), it is possible in principle to determine the molecular weights of polymorphic fragments in a sample by comparing them with monomorphic human bands in the same lane—so-called internal molecular-weight standards. These monomorphic fragments are expected to have undergone the same band shift, so they should provide an accurate internal ruler for measurement. (Note that the polymorphic fragments and the internal molecular-weight standards are visualized on separate hybridizations, but can be superimposed on one another, if the external molecular-weight standards are used to align the gels.)

In practice, however, the use of internal standards presents serious difficulties. Accurate size determination requires a number of internal standards. If band shifting caused all fragments to change their mobility by the same percentage, one would need only a single monomorphic fragment to determine the extent of shift. But band shifting appears to be more complex than that. Different regions of the gel shift by different amounts.

Little has been published on the nature of band shifting, on the number of monomorphic internal control bands needed for reliable correction, and on the accuracy and reproducibility of measurements made with such correction. For the present, several laboratories have decided against attempting quantitative corrections; samples that lie outside the match criterion because of apparent band shifting are declared to be "inconclusive." The committee urges further study of the problems associated with band shifting. Until testing laboratories have published adequate studies on the accuracy and reliability of such corrections, we recommend that they adopt the policy of declaring samples that show apparent band shifting to be "inconclusive."

The committee recommends that all measurement data be made readily available, including the computer-based images and records. Any analytical software for image processing or molecular-weight determination should also be readily available. All fragment sizes for both known and questioned samples should be clearly listed on the formal report of the testing laboratory.

Match Criteria

Current RFLP-based tests use VNTR probes that have dozens of closely spaced alleles. On the one hand, the high degree of polymorphism increases the power of the test to detect differences among persons. On the other hand, the large number of alleles increases the complexity of matching samples, because gels have little ability to resolve nearby alleles (which can differ by as little as 9 basepairs, so that, for practical purposes, the distribution of alleles can appear to be continuous).

Because of the limited resolution, two samples from a single person will often lead to slightly different measurements—e.g., 3.00 and 2.45 kilobases (kb) in one case, 3.03 and 2.40 kb in another. To decide whether two samples match, each laboratory must have a match criterion.[6] The match criterion should provide an objective and quantitative rule for deciding whether two patterns match—e.g., all fragments must lie within 2% of one another. When samples fall outside the match criterion, they should be declared to be "inconclusive" or "nonmatching."

The match criterion must be based on reproducibility studies that show the actual degree of variability observed when multiple samples from the same person are separately prepared and analyzed under typical forensic conditions. Some testing laboratories originally used matching rules that were based on the average spacing of fragment sizes in each region of the gel, rather than on actual studies of reproducibility. Other laboratories used purely visual matching criteria. Both are inadequate. Each testing laboratory must carry out its own reproducibility studies, because reproducibility varies among laboratories. The precise match criterion of each laboratory should be made freely available to all interested persons and should be stated in forensic reports.

The match criterion is also used in the calculation of allele frequencies. To determine the probability that a matching allele was found by chance, one counts the number of matching alleles in an appropriately chosen reference population. For the calculation to be valid, the same match criterion must be applied in screening the population databank and in comparing the forensic samples. Some testing laboratories originally used less stringent rules for declaring a match between forensic samples and more stringent rules for determining the frequency of matching alleles in the databank; the effect was an overstatement of the probability of obtaining a match by chance.

Some have advocated that testing laboratories, instead of using a match criterion, should report a likelihood ratio—the ratio of the probability that the measurements would have arisen if the samples came from the same person to the probability that they would have arisen if they came from different persons. No testing laboratories in the United States now use that approach. The committee recognizes its intellectual appeal, but recommends against it. Accuracy with it requires detailed information about the joint distribution of fragment positions, and it is not clear that information about a match could be understood easily by lay persons.

A laboratory's level of reproducibility can increase or decrease over time. Reproducibility should be measured not only when a laboratory first implements DNA typing, but continually on the basis of actual casework, as well as external proficiency testing (see Chapter 4). One easy way is to record the fragment measurements from the control samples of known DNA included on the membrane and regularly examine the variability in these measurements. A drawback of that approach is that the control pattern might become too well known to the examiners. A slight variation would eliminate the problem. Examiners would continue to use a fixed known control sample on every membrane, but would also be given a blind control sample as a bloodstain to analyze with each case. The latter sample would be randomly selected from a collection of a few dozen known samples. The examiners would not know its specific identity, but only a code number. They would compare the blind control sample against the known patterns, to determine whether it matched to the expected extent. Such an internal test of reproducibility would provide continuing internal measurement of a laboratory's reproducibility. It would likely be a powerful tool for quality control in a laboratory. For convenience, blind control samples could be distributed by a professional association or a private-sector firm. The committee recommends that testing laboratories adopt such a system for continuing measurement of reproducibility and that they regularly examine and report the results. Recommendations for mandating such testing systems are discussed in Chapter 4.

Retention of Sample

Scientifically, the best way to resolve ambiguity is often to repeat the experiment. The U.S. justice system guarantees opposing sides the right to have repeat experiments performed by experts of their choice, whenever that is possible. Accordingly, testing laboratories should measure DNA samples before analysis (with accurate devices, such as fluorometers, as well as with ethidium-stained "yield" gels) and should use only the quantity of DNA required for reliable Southern blot analysis. When they can, they should retain enough of a sample to repeat the entire analysis.

TECHNICAL ISSUES IN PCR-BASED METHODS

PCR is a relatively new technique in molecular biology, having come into common use in research laboratories only in the last 4 years. Although the basic exponential amplification procedure is well understood, many technical details are not, including why some primer pairs amplify much better than others, why some loci cause systematically unfaithful amplification, and why some assays are much more sensitive to variations in conditions. Nonetheless, it is an extremely powerful technique that holds great promise for forensic applications because of its great sensitivity and the potential of its use on degraded DNA.

We discuss here two broad categories of technical issues concerning PCR methods: issues related to the amplification step and issues related to the detection of amplified product.

Technical Issues Related to Amplification

Amplification Conditions

The quality and specificity of amplification with PCR depends on the amplification conditions: the amplification cycling program (temperatures, times, and number of cycles), the composition of the amplification mixture (e.g., primer, nucleotide, polymerase, and magnesium concentrations), and the amount and nature of the target DNA in the sample (single-stranded or double-stranded).[7] In some cases, results can vary among thermocyclers from different manufacturers, among thermocyclers of a single manufacturer, and even among different sample wells in a single machine. It is therefore essential that precise conditions be established for each typing system and that the system be thoroughly characterized for its sensitivity to variations in these conditions. If an assay yields spurious or confusing results under particular conditions, it may be necessary to prescribe strict condition limits or to discard the assay altogether. No PCR assay should be used until it has been rigorously characterized in this way.

Qualitative and Quantitative Fidelity

Ideally, PCR amplification products would faithfully represent the starting material in the sample—both qualitatively and quantitatively. But that is not always the case.

PCR amplification is known to result in misincorporation of nucleotides at the relatively low rate of less than one per 10,000 nucleotides per cycle.[8] Amplification is usually performed on a sample that contains a large number of molecules; if the misincorporation is random, the low frequency of random errors will not be detected in most systems and will pose no problem for the typing result. Difficulties arise for systems in which the misincorporation is not random. For example, DNA sequences that contain tandem repeat sequences—such as the dinucleotide $(CA)_n$ or some VNTRs—present serious problems. Apparently, the DNA polymerase can slip during amplification, introduce or delete copies of the repeat, and produce a heterogeneous collection of fragments, often making interpretation difficult. That drawback is unfortunate: such simple sequence repeats tend to be highly polymorphic in the human population and so would seem to be useful for forensics. Because there is no way to predict which PCR assays will be subject to this problem, each assay must be thoroughly characterized.

In some cases, PCR can be qualitatively faithful but quantitatively unfaithful, because some alleles amplify more efficiently than others. A sample might contain a 50:50 mixture of two alleles and yield an amplified product with a 90:10 ratio.[9] Differential amplification can arise through several mechanisms. It has been observed in the amplification of allelic products of different sizes (larger products tend to amplify less efficiently than shorter products) and in the amplification of sequences that differ significantly in GC content (because of differing denaturation efficiencies). In some cases, faithful amplification occurs at some temperatures and differential amplification at other temperatures.[8] The possibility of differential amplification needs to be addressed in the design and development of amplification protocols for each genetic-marker system. The safeguards to ensure that differential amplification does not occur should be defined and documented.

Quantitative analysis of mixed samples with PCR might be problematic. Suppose that PCR amplification reveals four alleles in a sample, and alleles 1 and 2 give a stronger signal than alleles 3 and 4. A conclusion that the two stronger alleles correspond to one contributor with genotype 1/2 and the two weaker alleles to a contributor with genotype 3/4 would be justified only if one had demonstrated that the amplification and detection process yielded signals that were directly proportional to the initial quantities of the alleles. If the locus were subject to differential amplification, the conclusion might be unjustified. This underscores the importance of characterizing possible differential amplification.

Ideally, primer pairs should amplify only the desired target locus. However, nonspecific amplification can be seen, if one amplifies for extended cycle numbers. Limits on the cycle number might be required as a safeguard against nonspecific products.

Amplification Inhibition

Some forensic samples contain factors that inhibit amplification, either by binding to the target DNA or by inhibiting the polymerase. In particular, amplification inhibition is often seen with DNA from older bloodstains. It can usually be remedied by re-extracting the DNA to remove the inhibiting factor, by diluting the offending DNA, or by increasing the concentration of polymerase. There is no evidence that any of those procedures affects typing adversely. Nevertheless, the nature of inhibiting factors and the mechanism of the inhibition effect deserve additional study. Each PCR system should be thoroughly characterized on a range of simulated and known forensic samples, to document any effect on reliability.

Contamination

One of the most serious concerns regarding PCR-based typing is contamination of evidence samples with other human DNA. PCR is not discriminating as to the source of the DNA it amplifies, and it can be exceedingly sensitive. Potentially, amplification of contaminant DNA could lead to spurious typing results. Three sorts of contamination can be identified, as set forth below; each has its own solutions.

• Mixed samples. Some evidence samples occur as mixtures, e.g., sexual-assault evidence, which often contains a mixture of semen and vaginal fluids. In mixed samples that contain semen, it is possible to extract the sperm DNA and the DNA of vaginal epithelial cells separately. That allows the genetic contribution of the male and female to be distinguished. However, there is one important caveat: if the sperm fraction shows a genotype that matches that of the victim, one cannot conclude that this represents the genotype of the perpetrator, inasmuch as it could be due to residual vaginal epithelial cells. The problem should disappear as PCR-based assays for more loci become available. For other mixtures, such separation is not possible. For example, it is not possible to separate the DNA contributed by different persons in mixed bloodstains or in sexual-assault samples that involve two or more perpetrators. Mixed samples are a reality of the forensic world that must be accommodated in interpretation and reconstruction. As a rule, mixed samples must be interpreted with great caution. Their interpretation should always be based on results from multiple PCR assays, so that one can check for consistency across various loci. Interpretations based on quantity can be particularly problematic—e.g., if one saw two alleles of strong intensity and two of weak intensity, it would be improper to assign the first pair to one contributor and the second pair to a second contributor, unless it had been firmly established that the system was quantitatively faithful under the conditions used.

• Contamination from handling in the field and laboratory. It is conceivable that DNA can be transferred to evidence samples or reaction solutions through handling, either from the person doing the handling or in transfer from other evidence samples. There are no hard data on the amounts of DNA transferred by physical contact, but there are anecdotal reports of experimenters who contaminated their PCR mixtures with their own DNA. It is difficult to assess the likelihood of this sort of contamination. Steps should be taken to minimize it, such as handling samples with gloves and preparing solutions and processing samples in separate areas. Contamination of solutions can be recognized with appropriate positive-control and blank-control amplifications, which should be used routinely. When a stain composed of blood, semen, or other biological material is analyzed with PCR, it is important to analyze unstained materials next to the stain with PCR as a control for contamination.

• PCR product carryover contamination. The most serious problem is contamination of evidence samples and reaction solutions with PCR products from prior amplifications. Such products can contain a target sequence at a concentration a million times greater, and even a relatively small quantity could swamp the correct signal from the evidence sample. Even the simple act of flipping the top of a plastic tube might aerosolize enough DNA to pose a problem.

Many research and diagnostic laboratories have been afflicted with the problem of PCR carryover. Contamination risks can be minimized by strict adherence to sterile technique; the use of separate work areas for sample processing, solution preparation, amplification, and type testing; the use of separate pipettes in each area (pipettes are a major source of carryover contamination); and maintenance of a one-way flow of materials from the evidence-storage area to the sample-preparation area to the type-testing area.

Those precautions focus primarily on preventing PCR carryover contamination. But it has become clear that carryover products from one PCR reaction to another must also be eliminated. One way is to use the nucleotide dUTP in place of dTTP in all PCR reactions.[10] PCR products made in this manner can be selectively destroyed by the enzyme uracil *N*-glycolase (UNG), which excises dUTP. Accordingly, all evidence samples would be treated before PCR amplification with UNG, to destroy contamination from previous PCR reactions. The method holds promise, although it has not yet been extensively tested in practice. Methods of detecting and preventing contamination from one PCR reaction to another in forensic laboratories are generally still in their early stages, and additional development should be encouraged.

As with contamination due to handling, carryover contamination can be signaled by the appearance of product in blank controls and of mixed or inappropriate types in samples and positive controls. Such controls should be used rigorously. Moreover, it should be remembered that the controls are useful for monitoring general contamination in the laboratory, not the accuracy of a particular experiment. If a blank control is positive in one experiment, it indicates a potential problem not just for that experiment, but for any experiment performed at about the same time—even in a laboratory contaminated with PCR carryover, blank controls do not necessarily become contaminated on every occasion. It will be wise to repeat all work with samples that have never been exposed to the PCR-typing laboratory.

In view of the problem of contamination due to handling and carryover, laboratories must incorporate contamination control into their standard operating procedures. And outbreaks of contamination and the steps taken to correct the problem should be documented.

One of the best safeguards against contamination is to have DNA typing independently performed in two laboratories, each starting with a piece of the unprocessed evidence sample. Given the inexpensiveness of typing, serious consideration should be given to independent replication of results—at least during the early stages of this technology.

Issues Related to Detection of Amplified Product

Variation after PCR amplification can be detected in several ways. The most popular detection schemes for nonforensic analyses are reverse dot hybridization, analysis of PCR products for size variation with gel electrophoresis, analysis of PCR products with gel electrophoresis after restriction-endonuclease digestion, analysis for the presence of amplification after use of allele-specific amplification primers, and analysis of the nucleotide sequence. No matter what detection scheme is used, contamination of the test sample with a second DNA sample or differential amplification of one allele in a sample that contains two alleles at the test locus can produce an error in typing. Differential amplification could result in the typing of a true heterozygote as a homozygote, or the low level of hybridization of the second allele could suggest the presence of a contaminating DNA in the test sample. A repeat PCR amplification and analysis might be successful and pinpoint a problem in the first amplification procedure. Besides those general problems, each detection format can entail its own technical problems.

Reverse Dot Hybridization

In forensic analysis today, the single PCR-based kit available uses reverse dot hybridization to detect variation at the HLA-DQα locus.[11] Reverse dot hybridization is based on a yes-no detection scheme. Theoretically, the absence of a signal in a dot means that the test allele is not present in the DNA sample, and a signal in a dot indicates the presence of the test allele. An intermediate signal in a dot—a signal that is considerably less intense than a second signal in the test hybridization—can result from a second DNA sample's contaminating the test sample, technical variation in the conditions of analysis (e.g., hybridization temperature), or true heterozygosity for the allele that produces the intermediate signal. Usually, such problems can be resolved through repeat experiments or by comparing results from a number of loci with many alleles (such loci could shed light on the nature of mixtures). If the DNA sample of a crime victim contains the allele in question, it would suggest contamination as the source. If repeat of the hybridization and washing procedures eliminates the intermediate signal, it would suggest its spurious nature. Rarely, the origin of an intermediate hybridization signal might remain unresolved; if the type of the sample is still questionable, the data should be discarded.

Other Detection Methods

When restriction-enzyme digestion of a PCR product is necessary to demonstrate alleles, the major pitfall is incomplete digestion. One can control the problem in the test sample by using a restriction enzyme for which a constant site that produces a fragment of constant size exists in the test fragment. For example, suppose that digestion of a 500-bp PCR fragment by *Hae*III yields a constant 300-bp fragment and either a 200-bp fragment or 120-bp and 80-bp fragments; a poor yield of the 300-bp fragment in a particular sample would suggest incomplete digestion of the PCR product.

Allele-specific amplification has been used in DNA diagnosis of genetic disease. Diagnosis can be based solely on absence of an amplification product, so it is a difficult technique to control adequately. Absence of a fragment can indicate either failure of the amplification procedure or absence of the allele in question from the test sample. If the former applies, a typing error would result. For this reason, the committee recommends that this method not be used for forensic analysis.

DNA sequence analysis of PCR products is commonly carried out either manually or with automation. Some ambiguity of nucleotide sequence at one or more positions is common and can signal DNA contamination or a technical problem in the analysis. Another important problem occurs when the DNA sequence of a fragment demonstrates variation at more than one position in the nucleotide sequence. Because both alleles at a locus are sequenced in the procedure, it is difficult to determine what fraction of the variation is contributed by each allele. For example, if heterozygosity is observed at two positions in a sequence, one cannot know, without further experimentation, whether one allele contains both variants or each allele contains one.

New methods of detection of PCR products will surely be devised. Well-controlled, extensive studies of the methods will be required before their use in forensic science, and the quality-assurance procedures described in Chapter 4 will be important to ensure their rigorous testing and reliability.

Use of Kits

One commercial kit for forensic PCR analysis has been marketed. Other such kits will probably be ready for commercial markets soon. The committee sees a potential for introduction of unreliable kits and the misuse of kits. The existence of a kit suggests ease of use and low chance of technical error. The committee believes that nonexpert laboratories will run a significant chance of error in using kits. We therefore recommend that a standing committee (discussed later in this chapter) consider the issue of regulatory approval of kits for commercial use in forensic DNA analysis. Even though no precedent exists for regulation of tests in forensic applications, we believe that it might be necessary for a government agency to test and approve kits for DNA analysis before their actual forensic use.

Prospects of PCR-Based Methods

PCR analysis has a number of desirable features for forensic applications. It requires very little DNA (less than for a Southern blot by a factor of 100-150) in the evidence sample. It is thus feasible to amplify dozens of loci. It generates a large quantity of relatively pure product that can be analyzed with much greater precision than Southern blots, even down to the nucleotide level. At the same time, it poses even more serious issues of proficiency, control, and technology transfer than RFLP typing.

In summary, it is well established that one can greatly amplify a locus with authenticity and that one can reliably detect alleles or sequence variation at the amplified locus with any of a number of techniques. PCR analysis is extremely powerful in medical technology, but it has not yet achieved full acceptance in the forensic setting. The theory of PCR analysis, even though it is the analysis of synthetic DNA, as opposed to the natural sample, is scientifically accepted and has been accepted by a number of courts. However, most forensic laboratories have invested their energy in development of RFLP technology and have left the development of forensic PCR technology to a few other laboratories. Thus, there is no broad base of experience in the use of the technique in identity testing.

Forensic PCR-based testing is now limited for the most part to analysis of genetic variation at the DQα locus in the HLA complex. Potential ambiguities in typing results cannot yet be checked by studying a number of other loci in the same DNA sample. That shortcoming will be rectified with the addition of new PCR markers for forensic analysis. However, it is clear that analysis of the DQα locus with PCR can often provide useful information during the investigative phase in the forensic setting.

In general, further experience should be gained with respect to PCR in identity testing. Information on the extent of the contamination problem in PCR analysis and the differential amplification of mixed samples needs to be further developed and published. A great deal of this information can be obtained when a number of polymorphic systems are available for PCR analysis. Ambiguous results obtained with a number of polymorphic markers will signal contamination or mixtures of DNA in a sample.

Quantification of PCR results needs to be explored, to make the results more reliable. Laboratories that gain experience with PCR should determine the relationship between cycle number and percentage of contaminating DNA easily detected for each system used. Control primers that amplify small amounts of DNA reliably and robustly need to be added to test amplifications. In general, information derived from new polymorphic loci under standardized conditions with easily quantifiable results or end points is needed. Considerable advances in the use of PCR in forensic analysis can be expected soon; the method has enormous promise.

NATIONAL COMMITTEE ON FORENSIC DNA TYPING

Forensic DNA typing is advancing rapidly. RFLP-based typing methods continue to be refined and improved, PCR typing methods are being used in some court cases, and other methods are being developed in scientific and commercial research laboratories. Typing methods will continue to be replaced with ever more sophisticated approaches for some time to come. These developments hold great promise for increasing the sensitivity and reliability of forensic DNA typing.

The rapidity of development creates a need to balance two competing societal objectives. On the one hand, new technologies should be made available quickly. On the other hand, forensic typing methods should not be used until their soundness is established both in principle and in practice. The problem involves both technology and technology transfer. Forensic DNA typing is drawing methods from the cutting edge of molecular genetics, but must apply them to quite different circumstances.

The committee believes that the field of forensic science would be best served by the creation of a National Committee on Forensic DNA Typing (NCFDT) to provide advice on scientific and technical issues as they arise. NCFDT would consist primarily of molecular geneticists, population geneticists, forensic scientists, and additional members knowledgeable in law and ethics. Its charges would be to provide guidance about the power and limitations of DNA typing methods, to identify potential problems and their solutions, to provide guidance about whether new technologies are ready for practical use in the forensic laboratory, and to provide advice concerning the regulation of kits for forensic DNA typing. In addition (as discussed in Chapter 3), NCFDT would provide advice on population genetics and statistical interpretation.

Such a committee could play a critical role in smoothing the acceptance of DNA typing technologies in the courtroom while ensuring their reliability. Although NCFDT would have no formal regulatory authority, we anticipate that substantial influence would derive from its stature and the quality of its advice, so that courts could look to its recommendations in making their decisions.

The present committee recommends that NCFDT be convened under the auspices of an appropriate government agency. Because its task is fundamentally scientific, we feel that the agency should be one whose primary mission is scientific, rather than related to law enforcement. To avoid any appearance of conflict of interest, an agency that uses forensic DNA typing itself would be unsuitable. Two excellent choices would be the National Institutes of Health (NIH) or the National Institute of Standards and Technology (NIST). NIH has extensive experience in molecular biology, population genetics, and laboratory practice. NIST has less direct experience in those fields, but has considerable experience in evaluating technologies. Regardless of which agency convenes NCFDT, we believe that the effort should have broad government support from NIST, NIH, the National Science Foundation, the National Institute of Justice, the Federal

Bureau of Investigation, and the State Justice Institute. NCFDT should also have broad support from the American Society of Crime Laboratory Directors, the Genetics Society of America, and the American Society of Human Genetics.

The creation of an expert advisory committee is a somewhat unusual step for forensic science. However, we feel that it is the appropriate way to ensure that the field can incorporate new developments promptly while maintaining high standards.

SUMMARY OF RECOMMENDATIONS

• Any new DNA typing method (or substantial variation on an existing method) must be rigorously characterized in both research and forensic settings, to determine the circumstances under which it will yield reliable results.

• DNA analysis in forensic science should be governed by the highest standards of scientific rigor, including the following requirements:

—Each DNA typing procedure must be completely described in a detailed, written laboratory protocol.
—Each DNA typing procedure requires objective and quantitative rules for identifying the pattern of a sample.
—Each DNA typing procedure requires a precise and objective matching rule for declaring whether two samples match.
—Potential artifacts should be identified by empirical testing, and scientific controls should be designed to serve as internal checks to test for the occurrence of artifacts.
—The limits of each DNA typing procedure should be understood, especially when the DNA sample is small, is a mixture of DNA from multiple sources, or is contaminated with interfering chemicals.
—Empirical characterization of a DNA typing procedure must be published in appropriate scientific journals.
—Before a new DNA typing procedure can be used, it must have not only a solid scientific foundation but also a solid base of experience.

• Regarding RFLP-based typing, the committee makes a number of technical recommendations, including specific recommendations about the choice of probes, the use of ethidium bromide in gels, controls for anomalous bands, measurement of fragment sizes, controls for band shifting, match criteria, and sample retention.

• Regarding PCR-based typing, the committee makes a number of technical recommendations, including recommendations for thorough characterization of each PCR assay for definition of the range of conditions under which it will perform reliably and for strict contamination measures and other control procedures.

• The committee strongly recommends the establishment of a National Committee on Forensic DNA Typing under the auspices of an appropriate government agency, such as NIH or NIST, to provide expert advice primarily on scientific and technical issues concerning forensic DNA typing.

REFERENCES

1. Balazs I, Baird M, Clyne M, Meade E. Human population genetic studies of five hypervariable DNA loci. Am J Hum Genet. 44:182-190, 1989.
2. Culliford BJ. Determination of phosphoglucomutase types in bloodstains. J Forensic Sci Sociol. 7:131-133, 1967.
3. Culliford BJ. The examination and typing of blood stains in the crime laboratory. Washington, D.C.: U.S. Government Printing Office, 1971.
4. Sensabaugh GF. Biochemical markers of individuality. pp. 338-415 in: Saferstein R, ed. Forensic science handbook. Englewood Cliffs, New Jersey: Prentice-Hall, 1982.
5. McNally L, Baird M, McElfresh K, Eisenberg A, Balazs I. Increased migration rate observed in DNA from evidentiary material precludes the use of sample mixing to resolve forensic cases of identity. Appl Theor Electrophoresis. 5:267-272, 1990.
6. Thompson WC, Ford S. The meaning of a match: sources of ambiguity in the interpretation of a DNA print in forensic DNA technology. pp. 93-152 in: Farley M, Harrington J, eds. Forensic DNA technology. Chelsea, Michigan: Lewis Publishing, 1991.
7. Arnheim N, Levenson CH. Polymerase chain reaction. C&E News 68:36-47, October 1, 1990.
8. Erlich HA, Gelfand D, Sninsky JJ. Recent advances in the polymerase chain reaction. Science. 252:1643-1651, 1991.
9. Comey CT, Jung JM, Budowle B. Use of formamide to improve amplification of HLA DQα sequences. Biotechniques. 10:60-61, 1991.
10. Longo MC, Berninger MS, Harley JL. Use of uracil DNA glycosylase to control carry-over contamination in polymerase chain reactions. Gene. 93:125, 1990.
11. Saiki RK, Walsh PS, Levenson CH, Erlich HA. Genetic analysis of amplified DNA with immobilized sequence-specific oligonucleotide probes. Proc Natl Acad Sci USA. 86:6230-6234, 1989.

3

DNA Typing: Statistical Basis for Interpretation

Can DNA typing uniquely identify the source of a sample? Because any two human genomes differ at about 3 million sites, no two persons (barring identical twins) have the same DNA sequence. Unique identification with DNA typing is therefore possible provided that enough sites of variation are examined.

However, the DNA typing systems used today examine only a few sites of variation and have only limited resolution for measuring the variability at each site. There is a chance that two persons might have DNA patterns (i.e., genetic types) that match at the small number of sites examined. Nonetheless, even with today's technology, which uses 3-5 loci, a match between two DNA patterns can be considered strong evidence that the two samples came from the same source.

Interpreting a DNA typing analysis requires a valid scientific method for estimating the probability that a random person might by chance have matched the forensic sample at the sites of DNA variation examined. A judge or jury could appropriately weigh the significance of a DNA match between a defendant and a forensic sample if told, for example, that "the pattern in the forensic sample occurs with a probability that is not known exactly, but is less than 1 in 1,000" (if the database that shows no match with the defendant's pattern is of size 1,000).

To say that two patterns match, without providing any scientifically valid estimate (or, at least, an upper bound) of the frequency with which such matches might occur by chance, is meaningless.

Substantial controversy has arisen concerning the methods for estimating the population frequencies of specific DNA typing patterns.[1-14] Questions have been raised about the adequacy of the population databases on which frequency estimates are based and about the role of racial and ethnic origin in frequency estimation. Some methods based on simple counting produce modest frequencies, whereas some methods based on assumptions about population structure can produce extreme frequencies. The difference can be striking: In one Manhattan murder investigation, the reported frequency estimates ranged from 1 in 500 to 1 in 739 billion, depending on how the statistical calculations were performed. In fact, both estimates were based on extreme assumptions (the first on counting matches in the databases, the second on multiplying *lower* bounds of each allele frequency). The discrepancy not only is a question of the weight to accord the evidence (which is traditionally left to a jury), but bears on the scientific validity of the alternative methods used for rendering estimates of the weight (which is a threshold question for admissibility).

In this chapter, we review the issues of population genetics that underlie the controversy and propose an approach for making frequency estimates that are independent of race and ethnic origin. This approach addresses the central purpose of DNA typing as a tool for the identification of persons.

ESTIMATING THE POPULATION FREQUENCY OF A DNA PATTERN

DNA "exclusions" are easy to interpret: if technical artifacts can be excluded, a nonmatch is definitive proof that two samples had different origins. But DNA "inclusions" cannot be interpreted without knowledge of how often a match might be expected to occur in the general population. Because of that fundamental asymmetry, although each new DNA typing method or marker can be used for investigation and exclusion as soon as its technical basis is secure, it cannot be interpreted with regard to inclusion until the population frequencies of the patterns have been established. We discuss the issues involved in estimating the frequency of a DNA pattern, consisting of pairs of alleles at each of several loci.

Estimating Frequencies of DNA Patterns by Counting

A standard way to estimate frequency is to count occurrences in a random sample of the appropriate population and then use classical statistical formulas to place upper and lower confidence limits on the estimate. Because estimates used in forensic science should avoid placing undue weight on incriminating evidence, an upper confidence limit of the frequency should be used in court. This is especially appropriate for forensic DNA typing, because any loss of power can be offset by studying additional loci.

To estimate the frequency of a particular DNA pattern, one might count the number of occurrences of the pattern in an appropriate random population sample. If the pattern occurred in 1 of 100 samples, the estimated frequency would be 1%, with an upper confidence limit of 4.7%. If the pattern occurred in 0 of 100 samples, the estimated frequency would be 0%, with an upper confidence limit of 3%. (The upper bound cited is the traditional 95% confidence limit, whose use implies that the true value has only a 5% chance of exceeding the upper bound.) Such estimates produced by straightforward counting have the virtue that they do not depend on theoretical assumptions, but simply on the sample's having been randomly drawn from the appropriate population. However, such estimates do not take advantage of the full potential of the genetic approach.

Estimating Frequencies of DNA Patterns with the Multiplication Rule (Product Rule)

In contrast, population frequencies often quoted for DNA typing analyses are based not on actual counting, but on theoretical models based on the principles of population genetics. Each matching allele is assumed to provide statistically independent evidence, and the frequencies of the individual alleles are multiplied together to calculate a frequency of the complete DNA pattern. Although a databank might contain only 500 people, multiplying the frequencies of enough separate events might result in an estimated frequency of their all occurring in a given person of 1 in a billion. Of course, the scientific validity of the multiplication rule depends on whether the events (i.e., the matches at each allele) are actually statistically independent.

From a statistical standpoint, the situation is analogous to estimating the proportion of blond, blue-eyed, fair-skinned people in Europe by separately counting the frequencies of people with blond hair, people with blue eyes, and people with fair skin and calculating their proportions. If a population survey of Europe showed that 1 of 10 people had blond hair, 1 of 10 had blue eyes, and 1 of 10 had fair skin, one would be wrong to multiply these frequencies to conclude that the frequency of people with all three traits was 1 in 1,000. Those traits tend to co-occur in Nordics, so the actual frequency of the combined description is probably higher than 1 in 1,000. In other words, the multiplication rule can produce an underestimate in this case, because the traits are correlated owing to population substructure—the traits have different frequencies in different population groups. Correlations between those traits might also be due to selection or conceivably to the action of some genes on all three traits. In any case, the example illustrates that correlations within subgroups—whatever their origin—bear on the procedures for estimating frequencies.

Unlike many of the technical aspects of DNA typing that are validated by daily use in hundreds of laboratories, the extraordinary population-frequency estimates sometimes reported for DNA typing do not arise in research or medical applications that would provide useful validation of the frequency of any particular person's DNA profile. Because it is impossible or impractical to draw a large enough population to test calculated frequencies for any particular DNA profile much below 1 in 1,000, there is not a sufficient body of empirical data on which to base a claim that such frequency calculations are reliable or valid per se. The assumption of independence must be strictly scrutinized and estimation procedures appropriately adjusted if possible. (The rarity of all the genotypes represented in the databank can be demonstrated by pairwise comparisons. Thus, in a recently reported analysis of the FBI database, no exactly matching pairs of profiles were found in five-locus DNA profiles, and the closest match was a single three-locus match among 7.6 million basepair comparisons.)[13]

The multiplication rule has been routinely applied to blood-group frequencies in the forensic setting. However, that situation is substantially different: Because conventional genetic markers are only modestly polymorphic (with the exception of human leukocyte antigen, HLA, which usually cannot be typed in forensic specimens), the multilocus genotype frequencies are often about 1 in 100. Such estimates have been tested by simple empirical counting. Pairwise comparisons of allele frequencies have not revealed any correlation across loci. Hence, the multiplication rule does not appear to lead to the risk of extrapolating beyond the available data for conventional markers. In contrast, highly polymorphic DNA markers exceed the informative power of protein markers, so multiplication leads to estimates that are less than the reciprocal of the size of the databases.

Validity of Multiplication Rule and Population Substructure

The multiplication rule is based on the assumption that the population does not contain subpopulations with distinct allele frequencies—that each individual's alleles constitute statistically independent random selections from a common gene pool. Under this assumption, the procedure for calculating the population frequency of a genotype is straightforward:

- Count the frequency of alleles. For each allele in the genotype, examine a random sample of the population and count the proportion of matching alleles—that is, alleles that would be declared to match according to the rule that is used for declaring matches in a forensic context. This step requires only the selection of a sample that is truly random with reference to the genetic type; it does not appeal to any theoretical models.

It is essential that the forensic matching rule be precise and objective—otherwise it would be impossible to apply it in calculating the proportion of individuals with matching alleles in the population databank. And it is essential that the same rule be applied to count frequencies in the population databank, because this is the only way to determine the proportion of random individuals that would have been declared to match in the forensic context. (In the context of forensic applications, an estimate of the probability of a match in DNA typing has been termed conservative if on the average it is larger than the actual one, so that any weight applied to the estimate would favor the suspect. Thus, some laboratories use a more conservative rule for counting population frequencies than for forensic matches—an acceptable approach, because it overestimates allele frequency. The converse would not be acceptable.)

• Calculate the frequency of the genotype at each locus. The frequency of a homozygous genotype a1/a1 is calculated to be p_{a1}^2, where p_{a1} denotes the frequency of allele a1. The frequency of a heterozygous genotype a1/a2 is calculated to be $2p_{a1}p_{a2}$, where p_{a1} and p_{a2} denote the frequencies of alleles a1 and a2. In both cases, the genotype frequency is calculated by simply multiplying the two allele frequencies, on the assumption that there is no statistical correlation between the allele inherited from one's father and the allele inherited from one's mother. The factor of 2 arises in the heterozygous case, because one must consider the case in which allele a1 was contributed by the father and allele a2 by the mother and vice versa: each of the two cases has probability $p_{a1}p_{a2}$. When there is no correlation between the two parental alleles, the locus is said to be in Hardy-Weinberg equilibrium. We should note that in forensic DNA typing, a slight modification is used in the case of apparently homozygous genotypes. When one observes only a single allele in a sample, one cannot be certain that the individual is a homozygote; it is always possible that a second allele has been missed for technical reasons. To be conservative, most forensic laboratories do not calculate the probability that the sample has two copies of the allele (which is p_{a1}^2), but rather the probability that the sample has at least one copy (which is $2p_{a1}$) leaving open the possibility of a second allele. We endorse this procedure.)

• Calculate the frequency of the complete multilocus genotype. The frequency of a complete genotype is calculated by multiplying the genotype frequencies at all the loci. As in the previous step, this calculation assumes that there is no correlation between genotypes at different loci; the absence of such correlation is called linkage equilibrium. (Some authors prefer to reserve the term linkage equilibrium for loci on the same chromosome and to use the term gametic phase equilibrium for loci on different chromosomes.) Suppose, for example, that a person has genotype a1/a2, b1/b2, c1/

c1. If a random sample of the appropriate population shows that the frequencies of a1, a2, b1, b2, and c1 are approximately 0.1, 0.2, 0.3, 0.1, and 0.2, respectively, then the population frequency of the genotype would be estimated to be [2(0.1)(0.2)][2(0.3)(0.1)][(0.2)(0.2)] = 0.000096, or about 1 in 10,417.

Again, the validity of the multiplication rule depends on the absence of population substructure, because only in this special case are the different alleles statistically uncorrelated with one another.

In a population that contains groups with characteristic allele frequencies, knowledge of one allele in a person's genotype might carry some information about the group to which the person belongs, and this in turn alters the statistical expectation for the other alleles in the genotype. For example, a person who has one allele that is common among Italians is more likely to be of Italian descent and is thus more likely to carry additional alleles that are common among Italians. The true genotype frequency is thus higher than would be predicted by applying the multiplication rule and using the average frequency in the entire population.

To illustrate the problem with a hypothetical example, suppose that a particular allele at a VNTR locus has a 1% frequency in the general population, but a 20% frequency in a specific subgroup. The frequency of homozygotes for the allele would be calculated to be 1 in 10,000 according to the allele frequency determined by sampling the general population, but would actually be 1 in 25 for the subgroup. That is a hypothetical and extreme example, but illustrates the potential effect of demography on gene frequency estimation.

Basis of Concern About Population Substructure

The key question underlying the use of the multiplication rule is whether actual populations have significant substructure for the loci used for forensic typing. This has provoked considerable debate among population geneticists: some have expressed serious concern about the possibility of significant substructure,[2,4,9,10] and others consider the likely degree of substructure not great enough to affect the calculations significantly.[1,3,6,8,11-13]

The population geneticists who urge caution make three points:

1. Population genetic studies show some substructure within racial groups for genetic variants, including protein polymorphisms, genetic diseases, and DNA polymorphisms. Thus, North American Caucasians, blacks, Hispanics, Asians, and Native Americans are not homogeneous groups. Rather, each group is an admixture of subgroups with somewhat different allele frequencies. Allele frequencies have not yet been homogenized, because people tend to mate within these groups.

2. For any particular genetic marker, the degree of subpopulation differentiation cannot be predicted, but must be determined empirically.

3. For the loci used for forensic typing, there have been too few empirical investigations of subpopulation differentiation.

In short, those population geneticists believe that the absence of substructure cannot be assumed, but must be proved empirically (see Lewontin and Hartl[10]). Other population geneticists, while recognizing the possibility or likelihood of population substructure, conclude that the evidence to date suggests that the effect on estimates of genotype frequencies are minimal (see Chakraborty and Kidd[12]). Recent empirical studies concerning VNTR loci[13, 14] (Weir, personal communication, 1991) detected no deviation from independence within or across loci. Moreover, pairwise comparisons of all five-locus DNA profiles in the FBI database showed no exact matches; the closest match was a single three-locus match among 7.6 million pairwise comparisons.[13] These studies are interpreted as indicating that multiplication of gene frequencies across loci does not lead to major inaccuracies in the calculation of genotype frequency—at least not for the specific polymorphic loci examined.

Although mindful of the controversy, the committee has chosen to assume for the sake of discussion that population substructure may exist and provide a method for estimating population frequencies in a manner that adequately accounts for it. Our decision is based on several considerations:

1. It is possible to provide conservative estimates of population frequency, without giving up the inherent power of DNA typing.

2. It is appropriate to prefer somewhat conservative numbers for forensic DNA typing, especially because the statistical power lost in this way can often be recovered through typing of additional loci, where required.

3. It is important to have a general approach that is applicable to any loci used for forensic typing. Recent empirical studies pertain only to the population genetics of the VNTR loci in current use. However, we expect forensic DNA typing to undergo much change over the next decade—including the introduction of different types of DNA polymorphisms, some of which might have different properties from the standpoint of population genetics.

4. It is desirable to provide a method for calculating population frequencies that is independent of the ethnic group of the subject.

Assessing Population Substructure Requires Direct Sampling of Ethnic Groups

How can one address the possibility of population substructure? In principle, one might consider three approaches: (1) carry out population studies on a large mixed population, such as a racial group, and use statistical tests to detect the presence of substructure; (2) derive theoretical principles that place bounds on the possible degree of population substructure; and (3) directly sample different groups and compare the observed allele frequencies. The third offers the soundest foundation for assessing population substructure, both for existing loci and for many new types of polymorphisms under development.

In principle, population substructure can be studied with statistical tests to examine deviations from Hardy-Weinberg equilibrium and linkage equilibrium. Such tests are not very useful in practice, however, because their statistical power is extremely low: even large and significant differences between subgroups will produce only slight deviations from Hardy-Weinberg expectations. Thus, the absence of such deviations does not provide powerful evidence of the absence of substructure (although the presence of such deviations provides strong evidence of substructure).

The correct way to detect genetic differentiation among subgroups is to sample the subgroups directly and to compare the frequencies. The following example is extreme and has not been observed in any U.S. population, but it illustrates the difference in power. Suppose that a population consists of two groups with different allele frequencies at a diallelic locus:

	A	a
Group I	0.5	0.5
Group II	0.9	0.1

If there is random mating within the groups, Hardy-Weinberg equilibrium within the groups will produce these genotype frequencies:

	AA	Aa	aa
Group I	0.25	0.50	0.25
Group II	0.81	0.18	0.01

Suppose that Group I is 90% of the population and Group II is 10%. In the overall population, the observed genotype frequencies will be

$$\begin{aligned} AA &= (0.9)(0.25) + (0.1)(0.81) = 0.306 \\ Aa &= (0.9)(0.50) + (0.1)(0.18) = 0.468 \\ aa &= (0.9)(0.25) + (0.1)(0.01) = 0.226 \end{aligned}$$

If we were unaware of the population substructure, what would we expect under Hardy-Weinberg equilibrium? The average allele frequencies will be

$$\begin{aligned} A &= (0.9)(0.5) + (0.1)(0.9) = 0.54 \\ a &= (0.9)(0.5) + (0.1)(0.1) = 0.46 \end{aligned}$$

which would correspond to the Hardy-Weinberg proportions of

AA = (0.54)(0.54) = 0.2916
Aa = 2(0.54)(0.46) = 0.4968
aa = (0.46)(0.46) = 0.2116

Even though there is substantial population substructure, the proportions do not differ greatly from Hardy-Weinberg expectation. In fact, one can show that detecting the population differentiation with the Hardy-Weinberg test would require a sample of nearly 1,200, whereas detecting it by direct examination of the subgroups would require a sample of only 22. In other words, the Hardy-Weinberg test is very weak for testing substructure.

The lack of statistical power to detect population substructure makes it difficult to detect genetic differentiation in a heterogeneous population. Direct sampling of subgroups is required, rather than examining samples from a large mixed population.

Similarly, population substructure cannot be predicted with certainty from theoretical considerations. Studies of population substructure for protein polymorphisms cannot be used to draw quantitative inferences concerning population substructure for VNTRs, because loci are expected to show different degrees of population differentiation that depend on such factors as mutation rate and selective advantage. Differences between races cannot be used to provide a meaningful upper bound on the variation within races. Contrary to common belief based on difference in skin color and hair form, studies have shown that the genetic diversity between subgroups within races is greater than the genetic variation between races.[15] Broadly, the results of the studies accord with the theory of genetic drift: the average allele frequency of a large population group (e.g., a racial group) is expected to drift more slowly than the allele frequencies of the smaller subpopulations that it comprises (e.g., ethnic subgroups).

In summary, population differentiation must be assessed through direct studies of allele frequencies in ethnic groups. Relatively few such studies have been published so far, but some are under way.[16] Clearly, additional such studies are desirable.

The Ceiling Principle: Accounting for Population Substructure

We describe here a practical and sound approach for accounting for possible population substructure: the ceiling principle.[9] It is based on the following observation: The multiplication rule will yield conservative estimates, even for a substructured population, provided that the allele frequencies used in the calculation exceed the allele frequencies in any of the population subgroups. Accordingly, applying the ceiling principle involves two steps: (1) For each allele at each locus, determine a *ceiling frequency* that is an upper bound for the allele frequency that is independent of the ethnic background of a subject; and (2) To calculate a genotype frequency, apply the multiplication rule, using the ceiling frequencies for the allele frequencies.

How should ceiling frequencies be determined? We must balance rigor and practicality. On the one hand, it is not enough to sample broad populations defined as "races" in the U.S. census (e.g., Hispanics), because of the possibility of substructure. On the other hand, it is not feasible or reasonable to sample every conceivable subpopulation in the world to obtain a guaranteed upper bound. The committee strongly recommends the following approach: Random samples of 100 persons should be drawn from each of 15-20 populations, each representing a group relatively homogeneous genetically; the largest frequency in any of these populations or 5%, whichever is larger, should be taken as the ceiling frequency. The reason for using 5% is discussed later.

We give a simplified example to illustrate the approach. Suppose that two loci have been studied in three population samples, with the following results:

	Population 1	Population 2	Population 3
Locus 1			
Allele a	1%	5%	11%
Allele b	5%	8%	10%
Locus 2			
Allele c	3%	4%	4%
Allele d	2%	15%	7%

For the genotype consisting of a/b at locus 1 and c/d at locus 2, the ceiling principle would assign ceiling values of 11% for allele a, 10% for allele b, 5% for allele c, and 15% for allele d and would apply the multiplication rule to yield a genotype frequency of [2(0.11)(0.10)][2(0.05)(0.15)] = 0.00033, or about 1 in 3,000. Note that the frequency used for allele c is 5%, rather than 4%, to reflect the recommended lower bound of 5% on allele frequencies. Because the calculation uses an upper bound for each allele frequency, it is believed to be conservative given the available data, even if there are correlations among alleles because of population substructure and even for persons of mixed or unknown ancestry. This is more conservative, and preferable, to taking the highest frequency calculated for any of the three populations.

The ceiling principle reflects a number of important scientific and policy considerations:

• The purpose of sampling various populations is to examine whether some alleles have considerably higher frequencies in particular subgroups than in the general population—presumably because of genetic drift. It is matches at such alleles that might be accorded too much evidentiary weight, if the general population frequency were used in calculating the probability of a match.

• Determining whether an allele has especially high frequency does not require a very large sample. A collection of 100 randomly chosen people provides a sample of 200 alleles, which is quite adequate for estimating allele frequencies.

• Genetically homogeneous populations from various regions of the world should be examined to determine the extent of variation in allele frequency. Ideally, the populations should span the range of ethnic groups that are represented in the United States—e.g., English, Germans, Italians, Russians, Navahos, Puerto Ricans, Chinese, Japanese, Vietnamese, and West Africans. Some populations will be easy to sample through arrangements with blood banks in the appropriate country; other populations might be studied by sampling recent immigrants to the United States. The choice and sampling of the 15-20 populations should be supervised by the National Committee on Forensic DNA Typing (NCFDT) described in Chapter 2.

We emphasize, however, that it is not necessary to be comprehensive. The goal is not to ensure that the ethnic background of every particular defendant is represented, but rather to define the likely range of allele frequency variation.

• Because only a limited number of populations can be sampled, it is necessary to make some allowance for unexamined populations. As usual, the problem is rare alleles. Genetic drift has the greatest proportional effect on rare alleles and may cause substantial variation in their frequency. Even if one sees allele frequencies of 1% in several ethnic populations, it is not safe to conclude that the frequency might not be five-fold higher in some subgroups.

To overcome this problem, we recommend that ceiling frequencies be 5% or higher. We selected this threshold because we concluded that allele frequency estimates that were substantially lower would not provide sufficiently reliable predictors for other, unsampled subgroups. Our reasoning was based on population genetic theory and computational results, and we aimed at accounting for the effects of sampling error and for genetic drift. The latter consideration was especially important, because it scales inversely with effective population size (i.e., small populations have larger drift) and because it accumulates over generations. The use of such a ceiling frequency would correspond to a lower bound of 5% on allele frequencies. Even if one observed allele frequencies of about 1%, one would guard against the possibility that the frequency in a subpopulation had drifted higher by using the lower bound of 5%. Thus, the lowest frequency attributable to any single locus would be 1/400 (1/20 × 1/20). In any case, it seems reasonable not to attach much greater weight to any single locus.

• The ceiling principle yields the same frequency for a genotype, regardless of the suspect's ethnic background, because the reported frequency represents a maximum for any possible ethnic heritage. Accordingly, the ethnic background of an individual suspect should be ignored in estimating the likelihood of a random match. The calculation is fair to suspects, because the estimated probabilities are likely to be conservative in their incriminating power.

Some legal commentators have pointed out that frequencies should properly be based on the population of possible perpetrators, rather than on the population to which a particular suspect belongs.[17,18] Although that argument is formally correct, practicalities often preclude use of that approach. Furthermore, the ceiling principle eliminates the need for investigating the perpetrator population, because it yields an upper bound to the frequency that would be obtained by that approach.

Some have proposed a Bayesian approach,[19-21] to the presentation of DNA evidence. However, this approach, focusing on likelihood ratios, does not avoid the kinds of population genetic problems discussed in this chapter. The committee has not tried to assess the relative merits of Bayesian and frequentist approaches, because, outside the field of paternity testing, no forensic laboratory in this country has, to our knowledge, used Bayesian methods to interpret the implications of DNA matches in criminal cases.

• Although the ceiling principle is a conservative approach, we feel that it is appropriate, because DNA typing is unique in that the forensic analyst has an essentially unlimited ability to adduce additional evidence. Whatever power is sacrificed by requiring conservative estimates can be regained by examining additional loci. (Although there could be cases in which the DNA sample is insufficient for typing additional loci with RFLPs, this limitation is likely to disappear with the eventual use of PCR.) A conservative approach imposes no fundamental limitation on the power of the technique.

DETERMINING ALLELE FREQUENCIES IN A POPULATION DATABANK

For forensic purposes, the frequency of an allele in a laboratory's databank should be calculated by counting the number of alleles that would be regarded as a match with the laboratory's forensic matching rule, which should be based on the empirical reproducibility of the system. This matching rule must account for both the quantitative reproducibility of forensic measurements in the testing laboratory and the quantitative reproducibility of the population measurements in the laboratory that generated the databank. In addition, the matching rule should reflect that one is making inter-gel comparisons, which are typically less precise than intra-gel comparisons.

The above approach is sometimes referred to as "floating bins," in that one counts the alleles that fall into a "bin" centered on the allele of interest. Most forensic laboratories in this country use the slightly different approach of "fixed bins":[22] One first aggregates alleles into a predetermined set of bins. Given an allele in a forensic case, one must then compute its frequency by adding the frequencies of all the bins that contain any alleles that fall within the window specified by the laboratory's forensic matching rule. (All bin frequencies must be added; it is not enough to take the largest of the bin frequencies.) This fixed-bin approach is acceptable and might be more convenient in some settings, because examiners need only consult a short table of bin frequencies, rather than search an entire databank.

IMPLICATIONS OF GENETIC CORRELATIONS AMONG RELATIVES

Because of the laws of Mendelian inheritance, the genotypes of biological relatives are much more similar than those of random individuals. Parent and child share exactly one identical allele at every locus, sibs share an average of one identical allele per locus, and grandparent and grandchild share an average of 0.5 identical allele per locus. (Here, identical refers to identity by descent from a common ancestor. Relatives can share additional alleles simply by chance.) These facts have important consequences for DNA typing:

- The genetic correlation between relatives makes it possible to carry out parentage and grandparentage testing. Paternity testing with DNA typing is already an active industry in the United States, and grandmaternity testing (with mitochondrial DNA, as well as nuclear genes) has been used in Argentina to reunite families with children who were abducted during the military dictatorship in the 1970s.[23,24] Relatedness testing involves a question analogous to that asked in identity testing: What is the chance that a randomly chosen person in the population would show the degree of relatedness expected of a relative? The same basic methods of population genetics apply, as discussed earlier.
- The ability to recognize relatedness poses a novel privacy issue for DNA databanks. Many states are starting to compile databanks that record patterns of DNA from convicted criminals, but not from other citizens, with the hope of identifying recidivists. When a biological sample is found at

the scene of a crime, its DNA pattern can be determined and compared with a databank. If the unidentified sample perfectly matches a sample in the convicted-criminal databank at enough loci, the probable perpetrator is likely to have been found. However, a different outcome could occur: the sample might match no entry perfectly, but match some entry at about one allele per locus. Depending on the number of loci studied, one could have a compelling case that the source of the sample was a first-degree relative (e.g., brother) of the convicted criminal whose entry was partially matched. (In practice, four loci would not suffice for this conclusion, but 10 might.) Such information could be sufficient to focus police attention on a few persons and might be enough to persuade a court to compel a blood sample that could be tested for exact match with the sample.

To put it succinctly, DNA databanks have the ability to point not just to individuals but to entire families—including relatives who have committed no crime. Clearly, this poses serious issues of privacy and fairness. As we discuss more fully later (Chapter 5), it is inappropriate, for reasons of privacy, to search databanks of DNA from convicted criminals in such a fashion. Such uses should be prevented both by limitations on the software for search and by statutory guarantees of privacy.

• Finally, the genetic correlation among relatives warrants caution in the statistical interpretation of DNA typing results. Our discussion above focused on the probability that a forensic sample would by chance match a person randomly chosen from the population. However, the probability that the forensic sample would match a relative of the person who left it is considerably greater than the probability that it would match a random person. Indeed, two sibs will often have matching genotypes at a locus—they have a 25% chance of inheriting the same pair of alleles from their parents and a 50% chance of inheriting one allele in common (which will result in identical genotypes if their other alleles happen to match by chance). Roughly speaking, the probability of a match at k loci will be approximately $(0.25 + 0.5p + 2p^2)^k$ in the general population, where p is the average chance that two alleles will match (i.e., the apparent homozygosity rate). Using p = 10% per locus for illustration, the probability that two sibs match at two loci would be about 10% and at four loci about 1%. Even for DNA profiles consisting entirely of very rare alleles (p~0%), the probability that two sibs will match at two loci is about 6% and at four loci about 0.3%. In short, the probability that two relatives will have matching genotypes is much greater than for two randomly chosen persons. Whenever there is a possibility that a suspect is not the perpetrator but is related to the perpetrator, this issue should be pointed out to the court. Relatives of a suspect could be excluded, of course, by testing their genotypes directly, provided that their DNA could be obtained.

IMPLICATIONS OF INCREASED POWER OF DNA TYPING COMPARED WITH CONVENTIONAL SEROLOGY

Questions about the population genetics of DNA markers remain open, but it is clear that the forensic scientist's discriminatory power has been substantially expanded with the advent of DNA markers. Indeed, forensic laboratories are routinely finding cases in which a suspect is included through conventional serology but later excluded through testing with DNA markers. The FBI reports, for example, that some 33% of suspects that match evidence samples according to conventional serology turn out to be excluded through DNA typing (J. W. Hicks, presentation to committee, 1990). Such outcomes represent a dramatic success of the new technology and often lead to the exoneration of innocent suspects.

LABORATORY ERROR RATES

Interpretation of DNA typing results depends not only on population genetics, but also on laboratory error. Two samples might show the same DNA pattern for two reasons: two persons have the same genotype at the loci studied, or the laboratory has made an error in sample handling, procedure, or interpretation. Coincidental identity and laboratory error are different phenomena, so the two cannot and should not be combined in a single estimate. However, both should be considered.

Early in the application of the DNA approach, results from nonblind proficiency studies suggested a high rate of false positives due to laboratory error. One commercial laboratory reported one false match in 50 samples in each of the first two blind proficiency tests conducted by the California Association of Crime Laboratory Directors (CACLD).[25] The error was attributed to incorrect sample loading in the first test and to mixing of DNA samples (because of reagent contamination) in the second. Another commercial laboratory reported no false positives in the two CACLD tests, but is reported to have made errors related to sample mixup in actual casework in *New York v. Neysmith*[26] and in the matter of a dead infant found in the Rock Creek area of Erie, Ill.[27] A third commercial laboratory made one error in 50 samples in the first CACLD test, but none in later blind trial testing. Estimates of laboratory errors in more recent practice are not available because of the lack of standardized proficiency testing.

Proficiency testing has also revealed important instances of false negatives. In the second CACLD test, the second laboratory cited failed to detect that two samples were 1:1 mixtures from two donors. Similarly, the first laboratory cited failed to detect several 1:1 mixtures and, in one case, reported that a stain from one person was a mixture. Those results raised serious questions about the reliability of interpretation of mixed samples.

Especially for a technology with high discriminatory power, such as DNA typing, laboratory error rates must be continually estimated in blind proficiency testing and must be disclosed to juries. For example, suppose the chance of a match due to two persons' having the same pattern were 1 in 1,000,000, but the laboratory had made one error in 500 tests. The jury should be told both results; both facts are relevant to a jury's determination.

Laboratory errors happen, even in the best laboratories and even when the analyst is certain that every precaution against error was taken. It is important to recognize that laboratory errors on proficiency tests do not necessarily reflect permanent probabilities of false-positive or false-negative results. One purpose of regular proficiency testing under standard case conditions is to evaluate whether and how laboratories have taken corrective action to reduce error rates. Nevertheless, a high error rate should be a matter of concern to judges and juries.

Reported error rates should be based on proficiency tests that are truly representative of case materials (with respect to sample quality, accompanying description, etc.). Tests based on pure blood samples would probably underestimate an error rate, and tests based primarily on rare and extremely difficult samples (which might be useful for improving practice) would probably overestimate. Although the CACLD proficiency test was less than ideal (being open, rather than blind, and not requiring reporting of size measurements), the materials appear to have been representative of standard casework.

TOWARD A FIRM FOUNDATION FOR STATISTICAL INTERPRETATION

Statistical interpretation of DNA typing evidence has probably yielded the greatest confusion and concern for the courts in the application of DNA to forensic science. Some courts have accepted the multiplication rule based on the grounds of allelic independence, others have used various ad hoc corrections to account for nonindependence, and still others have rejected probabilities altogether. Some courts have ruled that it is unnecessary even to test allelic independence, and others have ruled that allelic independence cannot be assumed without proof. The confusion is not surprising, inasmuch as the courts have little expertise in population genetics or statistics.

In reaching a recommendation on statistical interpretation of population frequencies, the committee balanced the following considerations:

- DNA typing should be able to provide virtually absolute individual identification (except in the case of identical twins), provided that enough loci are studied and that the population-genetics studies are developed with appropriate scientific care. The importance of this long-term goal justifies substantial investment in ensuring that the underlying population-genetics foundation is firm.

• Statistical testimony should be based on sound theoretical principles and empirical studies. Specifically, the validity of the multiplication rule in any application depends on the empirical degree of population differentiation for the loci involved. Adequate empirical data must be collected, and appropriate adjustments must be made to reflect the remaining uncertainties.

• It is feasible and important to estimate the degree of variability among populations to determine ceiling frequencies for forensic DNA markers and to evaluate the impact of population substructure on genotype frequencies estimated with the multiplication rule.

• Careful population genetics is especially important for the development and use of databanks of convicted-offender DNA patterns. Whereas the comparison of an evidence sample to a single suspect involves testing only one hypothesis, the comparison of a sample to an entire databank involves testing many alternative hypotheses. Special attention must thus be paid to the possibility of coincidental matches.

On the basis of those considerations, the committee reached conclusions, which now will be discussed.

Population Studies to Set Ceiling Frequencies

In view of the long-term importance of forensic DNA typing, the population-genetics foundation should be made as secure as possible. Accordingly, population studies should be promptly initiated to provide valid estimation of ceiling frequencies, as described above. Specifically, variation in allele frequencies should be examined in appropriately drawn random samples from various populations that are genetically relatively homogeneous. The selection, collection, and analysis of such samples should be overseen by the National Committee on Forensic DNA Typing (NCFDT) recommended in Chapter 2.

Given the effort involved in drawing appropriate population samples and the continuing need to type new markers as the technology evolves, the samples should be maintained as immortalized cell lines in a cell repository; that would make an unlimited supply of DNA available to all interested investigators. We note that preparation of immortalized cell lines through transformation of lymphoblasts with Epstein-Barr virus is routine and cost-effective. Transformation and storage can be handled as contract services offered by existing cell repositories, such as the NIH-supported repository in Camden, N.J.

Such a cell repository would be analogous to that of the international consortium Centre d'Etude du Polymorphisme Humain (CEPH)[28] created in 1983. It holds some 1,000 samples from 60 reference families, which are used for genetic mapping of human chromosomes. The cell lines have played an essential role in the development of the human genetic-linkage map. The existence of a common resource has also promoted standardization and quality control through the ability to recheck samples. (We should note that the CEPH families themselves are not appropriate for studying population frequencies, because they represent closely related people in a small number of families.)

Substantial benefits will accrue to forensic DNA typing through the availability of a reference collection that can be maintained at an existing facility like the ones at the Coriell Institute of Medical Research and the American Type Culture Collection. Although there is an initial investment in collecting, transforming, and storing cells, the cost will be more than repaid in the broad and continued availability of well-chosen samples for population studies of newly developed DNA typing systems and the ability of investigators to confirm independently the DNA typing that was done in another laboratory.

Reporting of Statistical Results

Until ceiling frequencies can be estimated from appropriate population studies, we recommend that estimates of population frequencies be based on existing data by applying conservative adjustments:

1. First, the testing laboratory should check to see that the observed multilocus genotype matches any sample in its population database. Assuming that it does not, it should report that the DNA pattern was compared to a database of N individuals from the population and no match was observed, indicating its rarity in the population. This simple statement based on the counting principle is readily understood by jurors and makes clear the size of the database being examined.

2. The testing laboratory should then calculate an estimated population frequency on the basis of a conservative modification of the ceiling principle, provided that population studies have been carried out in at least three major "races" (e.g., Caucasians, blacks, Hispanics, Asians, and Native Americans) and that statistical evaluation of Hardy-Weinberg equilibrium and linkage disequilibrium has been carried out (with methods that accurately incorporate the empirically determined reproducibility of band measurement) and no significant deviations were seen. The conservative calculation represents a reasonable effort to capture the actual power of DNA typing while reflecting the fact that the recommended population studies have not yet been undertaken. The calculation should be carried out as follows.

For each allele, a modified ceiling frequency should be determined by (1) calculating the 95% upper confidence limit for the allele frequency in each of the existing population samples and (2) using the largest of these values or 10%, whichever is larger. The use of the 95% upper confidence limit represents a pragmatic approach to recognize the uncertainties in current population sampling. The use of a lower bound of 10% (until data from ethnic population studies are available) is designed to address a remaining concern that populations might be substructured in unknown ways with unknown effect and the concern that the suspect might belong to a population not represented by existing databanks or a subpopulation within a heterogeneous group. We note that a 10% lower bound is recommended while awaiting the results of the population studies of ethnic groups, whereas a 5% lower bound will likely be appropriate afterwards. In the context of the discussion of the ceiling principle, the higher threshold reflects the greater uncertainty in using allele frequency estimates as predictors for unsampled subpopulations.

Once the ceiling for each allele is determined, the multiplication rule should be applied. The race of the suspect should be ignored in performing these calculations.

Regardless of the calculated frequency, an expert should—given with the relatively small number of loci used and the available population data—avoid assertions in court that a particular genotype is unique in the population. Finally, we recommend that the testing laboratory point out that reported population frequency, although it represents a reasonable scientific judgment based on available data, is an estimate derived from assumptions about the U.S. population that are being further investigated.

As an example, suppose that a suspect has genotype A1/A2, B1/B2 at loci A and B and that three U.S. populations have been sampled in the current "convenience sample" manner and typed for these loci. The likelihood of a match for this two-locus genotype would be estimated as follows:

	Population 1	Population 2	Population 3	Derived frequency
	750 persons	500 persons	200 persons	
Locus A				
Allele A1	0.003	0.013	0.042	Use 0.10
Allele A2	0.112	0.086	0.124	0.124 + 0.032 = 0.156[a]
Locus B				
Allele B1	0.004	0.007	0.014	Use 0.10
Allele B2	0.228	0.078	0.218	0.228 + 0.021 = 0.249[a]
Loci A and B combined	[2(0.10)(0.156)][2(0.10)(0.249)] = 0.001554			

[a]The upper 95% confidence limit is given by the formula $p + 1.96\sqrt{p(1-p)/N}$, where p is the observed frequency and N is the number of chromosomes studied.

A frequency of 0.001554 corresponds to about 1 in 644 persons. Addition of two loci with about the same information content would yield a four-locus genotype frequency of about 1 in 414,000 persons. Of course, if fewer than four loci were interpretable, as is common in forensic typing, the estimated genotype frequency would be much higher.

Significantly more statistical power for the same loci will be available when appropriate population studies have been carried out, because the availability of data based on a more rigorous sampling scheme will make it unnecessary to take an upper 95% confidence limit for each allele frequency nor to put such a conservative lower bound (0.10) on each allele frequency. Assuming that the population studies do not reveal significant substructure, the 5% lower bound recommended earlier should be used.

Finally, once appropriate population studies have been conducted and ceiling frequencies estimated under the auspices of NCFDT, population frequency estimates can be based on the ceiling principle (rather than the modified ceiling principle discussed above). Such calculations can never be perfect, but we believe that such a foundation will be sufficient for calculating frequencies that are prudently cautious—i.e., for calculating a lower limit of the frequency of a DNA pattern in the general population. In addition, new scientific techniques (e.g., minisatellite repeat codings[29]) are being and will be developed and might require re-examination by NCDFT of the statistical issues raised here.

Our recommendations represent an attempt to lay a firm foundation for DNA typing that will be able to support the increasing weight that will be placed on such evidence in the coming years. We recognize that a wide variety of methods for population genetics calculations have been used in previous cases—including some that are less conservative than the approach recommended here. We emphasize that our recommendations are not intended to question previous cases, but rather to chart the most prudent course for the future.

Openness of Population Databanks

Any population databank used to support forensic DNA typing should be openly available for reasonable scientific inspection. Presenting scientific conclusions in a criminal court is at least as serious as presenting scientific conclusions in an academic paper. According to long-standing and wise scientific tradition, the data underlying an important scientific conclusion must be freely available, so that others can evaluate the results and publish their own findings, whether in support or in disagreement. There is no excuse for secrecy concerning the raw data. Protective orders are inappropriate, except for those protecting individual's names and other identifying information, even for data that have not yet been published or for data claimed to be proprietary. If scientific evidence is not yet ready for both scientific scrutiny and public re-evaluation by others, it is not yet ready for court.

Reporting of Laboratory Error Rates

Laboratory error rates should be measured with appropriate proficiency tests and should play a role in the interpretation of results of forensic DNA typing. As discussed above, proficiency tests provide a measure of the false-positive and false-negative rates of a laboratory. Even in the best of laboratories, such rates are not zero.

A laboratory's overall rate of incorrect conclusions due to error should be reported with, but separately from, the probability of coincidental matches in the population. Both should be weighed in evaluating evidence.

SUMMARY OF RECOMMENDATIONS

Although mindful of the controversy concerning the population genetics of DNA markers, the committee has decided to assume that population substructure might exist for currently used DNA markers or for DNA markers that will be used in the future. The committee has sought to develop a recommendation on the statistical interpretation of DNA typing that is appropriately conservative, but at the same time takes advantage of the extraordinary power of individual identification provided by DNA typing. We have sought to develop a recommendation that is sufficiently robust, but is flexible enough to apply not only to markers now used, but also to markers that might be technically preferable in the future. We point out that in using conservative numbers in the interpretation of DNA typing results, any loss of statistical power is often offset through typing of additional loci. The committee seeks to eliminate the necessity to consider the ethnic background of a subject or of the group of potential perpetrators.

- As a basis for the interpretation of the statistical significance of DNA typing results, the committee recommends that blood samples be obtained from 100 randomly selected persons in each of 15-20 relatively homogeneous populations; that the DNA in lymphocytes from these blood samples be used to determine the frequencies of alleles currently tested in forensic applications; and that the lymphocytes be "immortalized" and preserved as a reference standard for determination of allele frequencies in tests applied in different laboratories or developed in the future. The collection of samples and their study should be overseen by a National Committee on Forensic DNA Typing.
- Sample collection and immortalization should be supported by feder-

al funds, in view of the benefits for law enforcement in general and for the convicted-offender databanks in particular.

• The ceiling principle should be used in applying the multiplication rule for estimating the frequency of particular DNA profiles. For each allele in a person's DNA pattern, the highest allele frequency found in any of the 15-20 populations or 5% (whichever is larger) should be used.

• In the interval (which should be short) while the reference samples are being collected, the significance of the findings of multilocus DNA typing should be presented in two ways: 1) If no match is found with any sample in a total databank of N persons (as will usually be the case), that should be stated, thus indicating the rarity of a random match. 2) In applying the multiplication rule, the 95% upper confidence limit of the frequency of each allele should be calculated for separate U.S. "racial" groups and the highest of these values or 10% (whichever is the larger) should be used. Data on at least three major "races" (e.g., Caucasians, blacks, Hispanics, Asians, and Native Americans) should be analyzed.

• Any population databank used to support DNA typing should be openly available for scientific inspection by parties to a legal case and by the scientific community.

• Laboratory error rates should be measured with appropriate proficiency tests and should play a role in the interpretation of results of forensic DNA typing.

REFERENCES

1. Devlin B, Risch N, Roeder K. No excess of homozygosity at loci used for DNA fingerprinting. Science. 249:1416-1420, 1990.
2. Cohen JE, Lynch M, Taylor CE, Green P, Lander ES. Forensic DNA tests and Hardy-Weinberg equilibrium. (Comment on Devlin et al. Science. 249:1416-1420, 1990.) Science. 253:1037-1039, 1991.
3. Devlin B, Risch N, Roeder K. (Response to Cohen et al. Science. 253:1037-1039, 1991). Science. 253:1039-1041, 1991.
4. Lander ES. Research on DNA typing catching up with courtroom application. (Invited Editorial.) Am J Hum Genet. 48:819-823, 1991.
5. Wooley JR. A response to Lander: The courtroom perspective. Am J Hum Genet. 49:892-893, 1991.
6. Caskey CT. Comments on DNA-based forensic analysis. (Response to Lander. Am J Hum Genet. 48:819, 1991.) Am J Hum Genet. 49:893-905, 1991.
7. Chakraborty R. Statistical interpretation of DNA-typing data. (Letter.) Am J Hum Genet. 49:895-897, 1991.
8. Daiger SP. DNA fingerprinting. (Letter.) Am J Hum Genet. 49:897, 1991.
9. Lander ES. Lander reply. (Letter.) Am J Hum Genet. 49:899-903, 1991.
10. Lewontin RC, Hartl DL. Population genetics in forensic DNA typing. Science. 254:1745-1750, 1991.
11. Chakraborty R, Daiger SP. Polymorphisms at VNTR loci suggest homogeneity of the white population of Utah. Hum Biol. 63:571-588, 1991.

12. Chakraborty R, Kidd K. The utility of DNA typing in forensic work. Science. 254:1735-1739, 1991.
13. Risch N, Devlin B. On the probability of matching DNA fingerprints. Science. 255:717-720, 1992.
14. Weir B. Independence of VNTR alleles defined as fixed bins. Genetics, in press.
15. Lewontin RC. The apportionment of human diversity. Evol Biol. 6:381-398, 1972.
16. Deka R, Chakraborty R, Ferrell RE. A population genetic study of six VNTR loci in three ethnically defined populations. Genomics. 11:83-92, 1991.
17. Cavalli-Sforza LL, Bodmer WF. The genetics of human populations. San Francisco: W.H. Freeman, 1971.
18. Lempert R. Some caveats concerning DNA as criminal identification evidence: with thanks to the Reverend Bayes. Cardozo Law Rev. 13:303-341, 1991.
19. Evett I, Werrett D, Pinchin R, Gill P. Bayesian analysis of single locus DNA profiles. Proceedings of the International Symposium on Human Identification 1989. Madison, Wisconsin: Promega Corp., 1990.
20. Berry DA. Influences using DNA profiling in forensic identification and paternity cases. Stat Sci. 6:175-205, 1991.
21. Berry DA, Evett IW, Pinchin R. Statistical inferences in crime investigation using DNA profiling. J Royal Stat Soc. [Series C - Applied Statistics], in press.
22. Budowle B, Giusti AM, Waye JS, Baechtel FS, Fourney RM, Adams DE, Presley LA, Deadman HA, Monson KL. Fixed-bin analysis for statistical evaluation of continuous distributions of allelic data from VNTR loci, for use in forensic comparisons. Am J Hum Genet. 48-841-855, 1991.
23. DiLonardo AM, Darlu P, Baur M, Orrego C, King MC. Human genetics and human rights: identifying the families of kidnapped children. Am J Forensic Med Pathol. 5:339-347, 1984.
24. King MC. An application of DNA sequencing to a human rights problem. Pp. 117-132 in: Friedmann T, ed. Molecular Genetic Medicine. Vol. 1. New York: Academic Press, 1991.
25. California Association of Crime Laboratory Directors, DNA Committee, Reports to the Board of Directors: 1, August 25, 1987; 2, November 19, 1987; 3, March 28, 1988; 4, May 18, 1988; 5, October 1, 1988; 6, October 1, 1988.
26. Lander ES. DNA fingerprinting on trial. (Commentary.) Nature. 339:501-505, 1989.
27. The fallibility of forensic DNA testing: of proficiency in public and private laboratories. Part I. The private sphere. Sci Sleuth Rev. 14(2):10, 1990.
28. Dausset J, Cann H, Cohen D, Lathrop M, Lalouel J-M, White R. Centre d'Etude du Polymorphisme Humain (CEPH): collaborative genetic mapping of the human genome. Genomics. 6:575-577, 1990.
29. Jeffreys A, MacLeod A, Tamaki K, Neil D, Monckton D. Minisatellite repeat coding as a digital approach to DNA typing. Nature. 354:204-209, 1991.

4

Ensuring High Standards

Critics and supporters of the forensic use of DNA typing agree that there has been a lack of standardization of practices and uniformly accepted methods for quality assurance. The lack is due largely to the rapid emergence of DNA typing and its introduction in the United States through the private sector.

As the technology developed in the United States, private laboratories using widely differing methods (single-locus RFLP, multilocus RFLP, and PCR) began to offer their services to law-enforcement agencies. During the same period, the FBI was developing its own RFLP method, with yet another restriction enzyme and different single-locus probes. Its method has become the one most widely (albeit not exclusively) used in public forensic-science laboratories, as a result of the FBI's national offering of free extensive training programs to forensic scientists. Each method has its own advantages and disadvantages, databanks, molecular-weight markers, match criteria, and reporting methods. In some courts, there have been differences of opinion as to the reliability, acceptability, and applicability of the various methods and particularly the degree of their specificity or discriminating power.

Regardless of the causes, practices in DNA typing vary and so do the educational backgrounds, training, and experience of the scientists and technicians who perform these tests, the internal and external proficiency testing conducted, the interpretation of results, and approaches to quality assurance.

It is not uncommon for an emerging technology to go without regulation until its importance and applicability are established. Indeed, the development of DNA typing technology has occurred without regulation of laboratories and their practices, public or private. The committee recognizes that standardization of practices in forensic laboratories in general is more problematic than in other laboratory settings: stated succinctly, forensic scientists have little or no control over the nature, condition, form, or amount of sample with which they must work. But it is now clear that DNA analytic methods are a most powerful adjunct to forensic science for personal identification and have immense benefit to the public—so powerful, so complex, and so important that some degree of standardization of laboratory procedures is necessary to assure the courts of high-quality results. DNA typing is capable, in principle, of an extremely low inherent rate of false-positive results, so the risk of error will come from poor laboratory practice or sample handling and labeling; and, because DNA typing is technical, a jury requires the assurance of laboratory competence in test results.

At issue, then, is how to achieve standardization of DNA typing laboratories in such a manner as to assure the courts and the public that results of DNA typing conducted and reported by a given laboratory are reliable, reproducible, and accurate.

DEFINING THE PRINCIPLES OF QUALITY ASSURANCE

Quality assurance can best be described as a documented system of activities or processes for the effective monitoring and verification of the quality of a work product (in this case, laboratory results). A comprehensive quality-assurance program must include elements that address education, training, and certification of personnel; specification and calibration of equipment and reagents; documentation and validation of analytic methods; use of appropriate standards and controls; sample handling procedures; proficiency testing; data interpretation and reporting; internal and external audits of all the above; and corrective actions to address deficiencies and weigh their importance for laboratory competence. An instructive example is *Guidelines for a Quality Assurance Program for DNA RFLP Analysis*, developed by the Technical Working Group on DNA Analysis Methods (TWGDAM).[1]

TWGDAM is a practitioners' group that comprises over 30 scientists who work in DNA typing in 24 state, local, and federal forensic laboratories in the United States and Canada. Its purpose is to assemble forensic scientists actively involved with DNA typing methods and have them discuss the methods being used, compare work and results, and share protocols. The FBI has subsidized and hosted the meetings and plays a central role in the activities of the group.

The TWGDAM guidelines established functions to be followed systematically in the RFLP typing procedure and cover many important aspects of the laboratory process. In addition, they provide documentation designed to ensure that DNA analysis is operating within the established performance criteria and provide a measure of the overall quality of results. TWGDAM has also published a more detailed description of the proficiency-testing portion of the quality-assurance guidelines.[2] The proficiency-testing guidelines for RFLP analysis describe the necessary elements of open and blind proficiency testing, including guidelines for documentation, review, and reporting of proficiency test results and deficiency and corrective actions.[3]

The TWGDAM guidelines, however, are just guidelines—they must be implemented in a formal, detailed quality-assurance program. The committee recommends that laboratories engaged in forensic DNA typing adhere to the TWGDAM guidelines for quality assurance and proficiency testing and implement them in formal programs. Although we admire the TWGDAM guidelines, we note that they do not go far enough in some ways. In some important respects, the technical recommendations contained in Chapters 2 and 3 of this report exceed the TWGDAM guidelines and should be seen as augmenting them.

TWGDAM is an excellent resource for forensic scientists and can play an important role to complement the National Committee on Forensic DNA Typing (NCFDT) recommended in Chapters 2 and 3. Whereas NCFDT should focus primarily on the key issues of scientific foundations and technology transfer, TWGDAM can serve as a forum for practitioners to discuss details of laboratory practice. Although TWGDAM was originally an FBI initiative, it should be restructured under broader auspices that represent the full range of forensic laboratories and societies.

Finally, on the subject of quality assessment and quality control, we note that the National Institute of Standards and Technology (NIST) has a program to test and develop a series of standards and controls for use in DNA typing. Because of its extensive experience in the development of standards, NIST can play a highly important role in standardization by developing lists of approved controls, equipment, and devices (including molecular-weight markers, monomorphic markers, cell lines, and typing kits).

POTENTIAL METHODS FOR ENSURING QUALITY

The principles of quality assurance discussed above must be enforced through appropriate mechanisms. Potential mechanisms include certification of individuals, accreditation of laboratories, mandatory licensing, and funding incentives contingent on adherence to standards. We discuss the advantages and disadvantages of these in turn.

Certification of Individuals

In the certification approach, a certifying body dictates that fulfillment of specified education, training, and experiential requirements be demonstrated by documentation and examination. An examination can also include a laboratory practicum. Individual certification has many advantages, but it is not adequate. A person does not perform DNA typing tests in isolation. A person's ability to produce high-quality results consistently depends heavily on the procedures, reagents, equipment, management, and attitudes in the work environment. It is impossible to separate a person from his or her organization and physical setting. In addition, personal certification can be expensive and usually requires funding support from the employing institution.

We recommend that the National Institute of Justice, given its interest in training, develop training programs in association with the American Society of Crime Laboratory Directors. Such a cooperative effort would allow continuity of candidate selection, training, examination, and certification.

Laboratory Accreditation

Accreditation is a more comprehensive approach to regulation. It requires that a laboratory demonstrate that its management, operations, individual personnel, procedures and equipment, physical plant and security, and personnel safety procedures all meet standards. Laboratory accreditation programs can be voluntary or mandatory. Although voluntary programs can have a positive effect, they suffer from the limitations that laboratories need not comply, that standard-setting need not be open to public scrutiny, and that accreditation might be contingent on membership in a professional organization. Accreditation programs required by federal or state law provide a greater level of assurance.

Licensing of Laboratories

Licensing involves vesting, by the federal government or a state government, of a regulatory body with the responsibility and authority to establish a series of requirements that a laboratory must satisfy if it is to be allowed to operate in a defined jurisdiction or to present evidence in its courts. The licensing approach does not suffer the disadvantage of being voluntary. State or federal laws can place sanctions on a laboratory that is not licensed by the specified body. A potential drawback is that the development of such a program can be time-consuming and expensive. In addition, licensing can be anticompetitive and can discourage innovation. In the case of state (as opposed to federal) regulation, there is the potential problem that different practices will be mandated in different jurisdictions, although this can be avoided if a few states take the lead and others follow suit by adopting similar regulations or recognize other states' licensing. The most efficient path seems to be for a federal licensing procedure to standardize the process for the nation and avoid state licensing, which can be restrictive and redundant. The federal license would be most cost-efficient and provide the field with a mechanism of quality assurance.

Funding Contingent On Adherence to Standards

The "stick" of licensing can be replaced by the "carrot" of funding. Such an approach would provide an incentive for adhering to standards, reduce costs for states and localities, and increase the number of laboratories able to afford DNA typing. A medical specialty laboratory typically spends $1,000-3,000 per annum for its licenses and accreditation. But the approach has numerous problems. The use of an equitable funding formula within state eligibility requirements might be problematic and controversial. Private laboratories would be unlikely to be covered (or even to want to be covered) by such a program. Expanding the number of practicing laboratories might decrease, not increase, the general standard of practice for a complex technology. Most important, the funding incentives alone do not provide an adequate guarantee of quality; they must be backed up by a regulatory program.

QUALITY ASSURANCE IN RELATED FIELDS

It is useful to consider the experience with various approaches in related fields. Medicine provides an appropriate analogy to forensic science, because it involves the application of sophisticated scientific methods to serious decisions that the practitioner must make on the basis of only partial information. Medicine has general mechanisms for setting standards of education and training of people, of laboratory practices and performance, and of quality assurance. The mechanisms involve both voluntary standard-setting by professional organizations and mandatory regulation through public licensing.

In pathology and laboratory medicine, the College of American Pathologists (CAP) is the dominant standard-setting organization. It has its roots in pathology, but has recently indicated its interest in working collaboratively with other laboratory-based professional groups toward establishing standard-setting for specialty laboratories. In medical genetics, the American Society of Human Genetics (ASHG) operates a personnel-certification program for clinicians and laboratory directors. ASHG recently agreed to provide expert consultants to CAP to establish laboratory certification in cytogenetics, biomedical genetics, and DNA-based diagnostics. In paternity testing, the American Association of Blood Banks (AABB) has developed laboratory-certification actions for protein typing methods and is establishing a set of standards for clinical DNA testing laboratories. AABB has also entered into discussions with CAP to codevelop proficiency testing for specialty laboratories. Additionally, the American Society for Histocompatibility and Immunogenetics has considered standards for histocompatibility testing, including DNA analysis.[4]

Mandatory government accreditation plays an essential role. Personnel certification and laboratory licensing are required in medicine, and formal requirements are set by state and federal regulatory bodies. Physicians, physician's assistants, medical technologists, laboratories, and others are licensed by states. Moreover, since 1967, a federal process of accreditation has been mandated for medical laboratories under the Clinical Laboratory Improvement Act (CLIA). In 1988, that process was expanded and restructured to enhance performance of laboratories involved in human DNA diagnostic programs. Officials responsible for implementing CLIA have recently initiated discussion with CAP for the purpose of codeveloping proficiency testing.

Medical accreditation programs have substantial force behind them. Government accreditation is mandated by law; private accreditation is often mandated as a component of government regulation; and both kinds of accreditation are required by third-party payers (e.g., insurance carriers) for payment.

INITIAL EFFORTS TOWARD ESTABLISHING STANDARDS IN FORENSIC DNA TYPING

Some initial efforts at developing accreditation and licensing standards for forensic DNA typing are already under way, as follows.

The American Society of Crime Laboratory Directors (ASCLD) is a professional organization of approximately 350 members that has represented forensic-science laboratory directors since 1975. Although most forensic laboratories in the United States are publicly funded and mandated by state or federal statutes to examine physical evidence and perform forensic testing, a voluntary laboratory-accreditation program has been in operation since 1985 through the auspices of the American Society of Crime Laboratory Directors-Laboratory Accreditation Board (ASCLD-LAB), which is an independently chartered organization affiliated with but separate from ASCLD. Some 77 forensic laboratories in the United States and in Australia are accredited by ASCLD-LAB. Such accreditation is available to all public and private forensic DNA laboratories, including ones that do not meet the membership requirements for ASCLD. There has, however, been no listing of laboratories that failed examination, had deficiencies, or were advised to discontinue providing services, and that constitutes a flaw in voluntary laboratory accreditation as carried out by ASCLD-LAB.

The New York state legislature has developed legislation to create a licensing program that was recommended by the Governor's Select Commission on DNA Typing, which was chaired by the state commissioner of criminal justice and consisted of lawyers, molecular biologists, forensic scientists, and population geneticists. It proposes that New York administer a licensing program with the advice of a scientific review board and a DNA advisory committee. Those bodies would consist of independent scientists in molecular biology and population genetics with no ties to forensic laboratories, through financial relationships, extensive collaboration, or provision of extensive testimony. This pioneering effort is consistent with the leadership role that New York has assumed in other laboratory testing—it is the national leader in the regulation of cytogenetics, for example. Both houses of the New York legislature passed a bill establishing a regulatory framework but it later was withdrawn. A similar bill has been reintroduced in the current legislative session.

A third approach is reflected in legislation recently introduced in the U.S. Congress "to provide financial assistance to state and local governments wishing to upgrade their crime laboratories with DNA genetic testing capability" contingent on their adherence to particular standards. The proposed law, entitled the DNA Proficiency Testing Act (H.R. 339, introduced by Rep. Frank Horton),[5] would promote increased quality assurance by providing grants from the Department of Justice (DOJ) Bureau of Justice Assistance to state and local forensic laboratories and mandating the FBI to publish "advisory standards for testing the proficiency of forensic laboratories" and to carry out certification for DNA proficiency-testing programs. The Bureau of Justice Assistance would make grants to state or local forensic laboratories. Within 6 months of enactment, the FBI would publish standards for testing the proficiency of laboratories that conduct DNA typing. The standards would "specify criteria to be applied to each procedure used by forensic laboratories to conduct analyses of DNA." The FBI would revise the standards from time to time, as necessary.

Under H.R. 339, the FBI could approve a DNA proficiency-testing program offered by a private organization, if the program satisfied the proficiency-testing standards and the testing laboratory were prepared to issue to each forensic laboratory that participated in the program a document that certified the participation and specified the period for which the proficiency test applied to the forensic laboratory. The FBI would have to withdraw approval for a DNA proficiency-testing program if the program or testing laboratory failed to satisfy the proficiency-testing standards.

In addition to those formal efforts for standardization now under consideration, the FBI has promoted standardization through various programs and initiatives.

- The FBI's Forensic Science Research and Training Center (FSRTC) in Quantico, Va., developed the analytic methods most commonly used or under development in forensic laboratories and promotes their acceptance through an excellent training program. Since January 1989, the FSRTC has provided training for over 200 forensic scientists from 74 state and local laboratories in the United States, South Korea, Turkey, and Portugal.
- The FBI has initiated, sponsored, and hosted the TWGDAM meetings at the FSRTC and published its guidelines.
- The FBI has proposed the establishment of a national DNA databank. The proposal has serious implications for the standardization of practices, in that its success would require participating forensic laboratories to be capable of producing nearly identical results on a given sample. That requirement would drive the forensic community to adopt some standardized analytic methods, including molecular-weight markers, probes, controls, and methods for allele measurement.
- The FBI provides educational and training opportunities (although it has not indicated an interest in developing and implementing personnel or laboratory testing programs for outside laboratories).

A REGULATORY PROGRAM FOR DNA TYPING

Components of a Suitable Program

In evaluating the relative merits of possible approaches, the committee developed a list of general requirements that a regulatory program adopted for the standardization of DNA typing technology should include. The ideal program would contain mechanisms to ensure that:

- Individual analysts have education, training, and experience commensurate with the analysis performed and testimony provided.
- Analysts have a thorough understanding of the principles, use, and limitations of methods and procedures applied to the tests performed.
- Analysts successfully complete periodic proficiency tests and their equipment and procedures meet specified criteria.
- Reagents and equipment are properly maintained and monitored.
- Procedures used are generally accepted in the field and supported by published, reviewed data that were gathered and recorded in a scientific manner.
- Appropriate controls are specified in procedures and are used.
- New technical procedures are thoroughly tested to demonstrate their

efficacy and reliability for examining evidence material before being implemented in casework.

- Clearly written and well-understood procedures exist for handling and preserving the integrity of evidence, for laboratory safety, and for laboratory security.
- Each laboratory participates in a program of external proficiency testing that periodically measures the capability of its analysts and the reliability of its analytic results.
- Case records—such as notes, worksheets, autoradiographs, and population databanks—and other data or records that support examiners' conclusions are prepared, retained by the laboratory, and made available for inspection on court order after review of the reasonableness of a request.
- Redundancy of programs is avoided, so that unnecessary duplication of effort and costs can be eliminated.
- The program is widely accepted by the forensic-science community.
- The program is applicable to federal, state, local, and private laboratories.
- The program is enforceable—i.e., that failure to meet its requirements will prevent a laboratory from continuing to perform DNA typing tests until compliance is demonstrated.
- The program can be implemented within a relatively short time.
- The program involves appropriate experts in forensic science, molecular biology, and population genetics.

The Role of Professional Organizations

One guarantee of high quality is the presence of an active professional organization that is committed and able to enforce standards. Historically, the professional societies in forensic science have not played a very active role—certainly much less than medical societies. That has been due to a variety of circumstances, including the fact that accreditation and proficiency testing can be expensive and can lead to serious repercussions for laboratories with poor performance. Voluntary programs have few incentives and offer relatively little credibility for participating laboratories. Moreover, courts have not required certification, accreditation, or proficiency testing for admissibility of evidence. Together, those factors have worked against the development of rigorous accreditation programs.

Recently, however, ASCLD-LAB has shown a substantial interest in assuming an active role. At the annual meeting of ASCLD and ASCLD-LAB in September 1990, the boards of both organizations passed—with near unanimity—a resolution to expand requirements for accreditation of forensic-science laboratories engaged in DNA typing, including mandatory proficiency testing at regular intervals.[6,7]

Specifically, new standards require:

- Mandatory participation in an external proficiency-testing program administered by a contractor that meets rigid specifications for adequacy and reliability (it is envisioned that two sets of tests will be required each year).
- The requirement that each examiner in a laboratory take proficiency tests and submit results independently.
- Review of test results by ASCLD-LAB.
- Inclusion in the board's DNA Advisory Committee of both forensic DNA scientists and leading experts in nonforensic DNA technology, to provide guidance and advice in forensic DNA testing.
- Periodic on-site inspections of accredited laboratories.
- Documentation of conduct of internal, blind, and open proficiency testing.
- Adherence to TWGDAM guidelines for quality assurance and proficiency testing (endorsed by ASCLD and ASCLD-LAB).
- A mechanism to revoke or suspend accreditation on documentation of unsatisfactory performance.

ASCLD has unanimously recommended that any forensic laboratory engaged in DNA typing or actual case materials be accredited by ASCLD-LAB for the typing method used. The committee supports the effort by ASCLD and ASCLD-LAB, because it holds the promise of rallying the profession to enforce high standards on its members. Of course, it remains for ASCLD-LAB to demonstrate that it will actively discharge the role of accreditation and mandatory proficiency testing. Any accreditation program would need to be advised by persons without ties to forensic laboratories that are engaged in DNA typing, in order to assure the public—especially defendants—that it is independent and objective. To that end, program leadership should consist primarily of persons who have no financial stake in the practice of forensic DNA testing.

The Role of Government

Voluntary accreditation programs are not enough. Because professional organizations, such as ASCLD-LAB, lack regulatory authority, forensic laboratories could avoid accreditation and still offer DNA typing evidence in criminal proceedings. In view of the important public-policy goal that this powerful technology be practiced only at the highest standard, compliance with high standards must be mandatory. Two approaches should be used to accomplish this, as set forth below.

First, courts should require that a proponent of DNA typing evidence have appropriate accreditation—including demonstration of external, blind proficiency testing (as well as other accreditation that might be mandated by government or come to be generally accepted in the profession)—for its evidence to be admissible. There is strong legal foundation for such a position. As a number of courts have correctly recognized, the admissibility of scientific evidence depends not just on a technology's being sound in principle, but on the testing laboratory's having applied it in the case at hand according to generally accepted standards. Courts should view the absence of appropriate accreditation as constituting a prima facie case that the laboratory has not complied with generally accepted standards. Until accreditation programs are fully implemented, there will be a period during which some laboratories will not have completed the accreditation process. In the interim, courts should require forensic laboratories at least to demonstrate that they are effectively in compliance with the requirements for accreditation as outlined by TWGDAM and by this report; that would be taken as meeting generally accepted standards of practice.

Second, the federal government should adopt legislation requiring accreditation of all forensic laboratories engaged in DNA typing. The committee recommends the following approach:

- Establishing mandatory accreditation should be a responsibility of the Department of Health and Human Services (DHHS), in consultation with the Department of Justice (DOJ). DHHS is the appropriate agency, because it has extensive experience in the regulation of clinical laboratories through programs under the Clinical Laboratory Improvement Act and has extensive expertise in molecular genetics through the National Institutes of Health. DOJ must be involved, because the task is important for law enforcement. However, the committee feels that primary responsibility should rest with DHHS for two reasons. First, DOJ lacks expertise in quality assurance and quality control and in molecular or population genetics. Second, DOJ may be perceived as an advocate for application of the technology. Oversight by DOJ may not be perceived as providing adequate assurance to the public or to a defendant facing prosecution by DOJ or affiliated agencies.
- DHHS, in consultation with DOJ, should contract with an organization to create and administer an appropriate laboratory-accreditation program, including proficiency testing. Preferably, the organization would be a private professional group; that would avoid the creation of additional government bureaucracy. One choice would be ASCLD-LAB, provided that it followed through on its announced intentions to assume a more active standard-setting role and demonstrated rigorous independence from the laboratories that it would regulate. Another choice would be CAP, which has unparalleled experience in administering such laboratory-accreditation programs in a wide variety of fields. Alternatively, the organization could be a government agency, provided that it were not itself engaged in operat-

ing forensic laboratories or closely tied to such a laboratory. Thus, NIST might be appropriate, whereas the FBI and TWGDAM (suggested for this role by H.R. 339) would not be.

- Mandatory accreditation falls within the interstate commerce clause, inasmuch as the standard of practice of forensic DNA typing laboratories profoundly affects citizens from states other than those of the laboratories. Thus, even state forensic laboratories would be covered.
- With the aid of an outside panel of experts, DHHS, in consultation with DOJ, should periodically determine whether the accrediting organization and the accreditation program are performing satisfactorily and, if not, select a new organization.

As in medicine, federal legislation should not preclude additional state licensing, but it might avoid the need for duplication of equivalent licensing programs at multiple levels.

The committee considers mandatory accreditation to be essential for ensuring high standards in DNA typing.

Support for Education, Training, and Research

In the long term, high quality in a field depends on first-rate programs of education, training, and research. Such programs are crucial for developing the necessary foundation of people and knowledge. The National Institute of Justice (NIJ) is the appropriate organization for such efforts and for assisting in education, training, and research. It is our opinion that NIJ has been given inadequate financial resources to meet these needs and that the level of support should be re-evaluated.

SUMMARY OF RECOMMENDATIONS

- Although standardization of forensic practice is somewhat problematic because of the nature of the samples, DNA typing is such a powerful and complex technology that some degree of standardization is necessary to ensure high standards.
- Each forensic-science laboratory engaged in DNA typing must have a formal, detailed quality-assurance and quality-control program to monitor work, on both an individual and a laboratory-wide basis.
- The TWGDAM guidelines for a quality-assurance program for DNA RFLP analysis are an excellent starting point for a quality-assurance program, which should be supplemented by the additional technical recommendations made in Chapters 2 and 3 of this report.
- TWGDAM should continue to function, playing a role complementary to that of the National Committee on Forensic DNA Typing. To increase

its effectiveness, TWGDAM should include more independent technical experts from outside the forensic community and should not be closely tied to any forensic laboratory.

• Quality-assurance programs in individual laboratories alone are insufficient to ensure high standards. External mechanisms are needed, to ensure adherence to the practices of quality assurance. Potential mechanisms include individual certification, laboratory accreditation, and state or federal regulation.

• One of the best guarantees of high quality is the presence of an active professional-organization committee that is able to enforce standards. Although professional societies in forensic science have historically not played an active role, ASCLD and ASCLD-LAB recently have shown substantial interest in enforcing quality by expanding the ASCLD-LAB accreditation program to include mandatory proficiency testing. ASCLD-LAB must demonstrate that it will actively discharge this role.

• Because private professional organizations lack the regulatory authority to require accreditation, further means are needed to ensure compliance with appropriate standards.

• Courts should require that proponents of DNA typing evidence have proper accreditation for each DNA typing method used. Lack of accreditation should be considered to constitute a prima facie case that a laboratory has not complied with generally accepted standards.

• In view of the compelling public interest in ensuring high standards for DNA typing, the federal government should enact legislation to establish a mandatory accreditation program. DHHS, in consultation with DOJ, should be assigned responsibility for engaging an appropriate organization to develop and administer the program. Possible choices for such an organization include ASCLD-LAB, CAP, and NIST.

• NIJ does not appear to receive adequate funds to support education, training, and research in this field properly. The level of funding should be re-evaluated and increased appropriately.

REFERENCES

1. Technical Working Group on DNA Analysis Methods (TWGDAM). Guidelines for a quality assurance program for DNA restriction fragment length polymorphism analysis. Crime Lab Dig. 16(2):40-59, 1989.
2. Technical Working Group on DNA Analysis Methods (TWGDAM). Guidelines for a proficiency testing program for DNA restriction fragment length polymorphism analysis. Crime Lab Dig. 17(3):50-60, 1990.
3. Technical Working Group on DNA Analysis Methods (TWGDAM). Statement of the Working Group on Statistical Standards for DNA Analysis. Crime Lab Dig. 17(3):53-58, 1990.

4. American Society for Histocompatibility and Immunogenetics. Standards for histocompatibility testing. ASHI Quart. 14(1), Winter-Spring, 1990.
5. H.R. 339, DNA Proficiency Testing Act of 1991, 102nd Congress, 1st Session. January 8, 1991.
6. Resolution of the ASCLD-LAB Delegate Assembly, Sept. 26, 1990.
7. ASCLD Board of Directors Resolution of Support, Sept. 27, 1990.

5

Forensic DNA Databanks and Privacy of Information

DNA typing in the criminal-justice system has so far been used primarily for direct comparison of DNA profiles of evidence samples with profiles of samples from known suspects. However, that application constitutes only the tip of the iceberg of potential law-enforcement applications. If DNA profiles of samples from a population were stored in computer databanks (databases), DNA typing could be applied in crimes without suspects. Investigators could compare DNA profiles of biological evidence samples with a databank to search for suspects.

In many respects, the situation is analogous to that of latent fingerprints. Originally, latent fingerprints were used for comparing crime-scene evidence with known suspects. With the development of the Automated Fingerprint Identification Systems (AFIS) in the last decade, the investigative use of fingerprints has dramatically expanded. Forensic scientists can enter an unidentified latent-fingerprint pattern into the system and within minutes compare it with millions of people's patterns contained in a computer file. In its short history, automated fingerprint analysis has been credited with solving tens of thousands of crimes.[1]

This chapter examines whether similar databanks of DNA profiles should be created and, if so, how and when.

COMPARISON OF DNA PROFILES AND LATENT FINGERPRINTS

To identify key issues pertinent to the establishment of DNA databanks, it is instructive to compare DNA profiles and latent fingerprints.

• Latent fingerprints are found at crime scenes much more commonly than are body fluids that contain DNA. Latent-fingerprint analysis can be useful in a wide range of crimes, including many murders, rapes, assaults, robberies, and burglaries. However, the probative value of latent fingerprints is often limited to establishing that a suspect was present at a location—and that does not automatically imply guilt. DNA analysis will be useful in more limited settings. DNA analysis will be useful primarily in rapes (because semen is often recovered) and murders (those in which either the perpetrator's blood was spilled at the crime scene or the victim's blood stained the perpetrator's personal effects—only the former will assist in identifying an unknown suspect). Where it exists, DNA evidence will often be more probative than fingerprints, in that the presence of body fluids is harder to attribute to innocuous causes. That is especially true in rape cases, in which positive identification of semen in the vagina is virtual proof of intercourse (although it leaves open the issue of whether it was consensual). Consequently, the potential utility of a DNA profile databank must be evaluated in terms of the particular crimes to which it is primarily suited.

• Fingerprints have a defined physical pattern independent of the method of visualization, whereas DNA profiles are derived patterns that can be constructed with various protocols (e.g., different restriction enzymes to cut the DNA and different probes to examine different loci) that produce completely different patterns that cannot be readily interconverted. The advance of DNA technology will see the development of new protocols that offer technical advantages but produce different and incompatible patterns.

In a sense, current DNA profiles can be thought of as extremely small bits of a person's fingerprints on all or some of the fingers. Different methods look at different fingers or different locations on a finger. Only when DNA technology is capable of sequencing the entire three billion basepairs of a person's genome could a DNA pattern be considered to be as constant and complete as a fingerprint pattern. Consequently, the development of DNA databanks is tied to the standardization of methods. A national DNA profile databank can function only if participating laboratories agree on standardized methods. However, the creation of a databank with current methods could discourage the conversion to newer, cheaper, and more powerful methods.

• The amount of information provided by latent fingerprints in an evidence sample is essentially fixed—it depends primarily on the portion of the finger(s) or palm found—and the forensic scientist uses all of it. DNA typing of an evidence sample yields information in an amount determined by the number of loci studied, so the forensic scientist has substantial control over the amount of information to be obtained from a sample. Consequently, the creation of a DNA profile databank would require decisions about the extent of the DNA profile to be recorded.

• Fingerprints are more highly individualized than DNA profiles based on the RFLP technology being used in forensic laboratories. Consequently, a match between an evidence sample and an entry in a DNA profile databank should not automatically lead to the assumption of identity, but should be confirmed by the examination of additional loci that are not in the databank.

• Obtaining an inked fingerprint from a person is much less intrusive, costly, and difficult than drawing a blood sample for DNA typing.

• Collection of fingerprints from known persons is inexpensive and relatively easily accomplished by someone with minimal technical background and training. In contrast, development of a DNA profile from a blood sample is time-consuming and expensive and requires extensive education, training, and quality-assurance measures. Consequently, the number of people who can be included in a DNA profile databank might be limited by economic considerations. Categories of persons to include must be selected with due consideration of costs and benefits.

• The computer technology required for an automated fingerprint identification system is sophisticated and complex. Fingerprints are complicated geometric patterns, and the computer must store, recognize, and search for complex and variable patterns of ridges and minutiae in the millions of prints on file. Several commercially available but expensive computer systems are in use around the world. In contrast, the computer technology required for DNA databanks is relatively simple. Because DNA profiles can be reduced to a list of genetic types (i.e., a list of numbers), DNA profile repositories can use relatively simple and inexpensive software and hardware. Consequently, computer requirements should not pose a serious problem in the development of DNA profile databanks.

• Fingerprints provide no information about a person other than identity. DNA typing can, in principle, also provide personal information—concerning medical characteristics, physical traits, and relatedness—that carries with it risks of discrimination. Consequently, DNA typing raises considerably greater issues of privacy than does ordinary fingerprinting.

In short, ordinary fingerprints and DNA profiles differ substantially in ways that bear on the creation and design of a national DNA profile databank.

CONFIDENTIALITY AND SECURITY

Confidentiality and security of DNA-related information are especially important and difficult issues, because we are in the midst of two extraordinary technological revolutions that show no signs of abating: in molecular biology, which is yielding an explosion of information about human genetics, and in computer technology, which is moving toward national and international networks connecting growing information resources.

Molecular geneticists are rapidly developing the ability to diagnose a wide variety of inherited traits and medical conditions. The list already includes simply inherited traits, such as cystic fibrosis, Huntington's disease, and some inherited cancers. In the future, the list might grow to include more common medical conditions, such as heart disease, diabetes, hypertension, and Alzheimer's disease. Some observers even suggest that the list could include such traits as predispositions to alcoholism, learning disabilities, and other behavioral traits (although the degree of genetic influence on these traits remains uncertain).

Obviously, such information could lead to discrimination by insurance companies, employers, or others against people with particular traits. In general, the committee feels that DNA profile databanks should avoid the use of loci associated with traits or diseases. That avoidance is the best guarantee against misuse of such information. Current forensic RFLP typing markers are not known to be associated with particular traits or medical conditions, but they might be in the future. Current PCR typing uses the HLA DQα locus, which is in a gene that controls many important immunological functions and is associated with diseases.

Even simple information about identity requires confidentiality. Just as fingerprint files can be misused, DNA profile identification information could be misused to search and correlate criminal-record databanks or medical-record databanks. Computer storage of information increases the possibilities for misuse. For example, addresses, telephone numbers, social security numbers, credit ratings, range of incomes, demographic categories, and information on hobbies are currently available for many of the citizens in our society from various distributed computerized data sources. Such data can be obtained directly through access to specific sources, such as credit-rating services, or through statistical disclosure.[2] "Statistical disclosure" refers to the ability of a user to derive an estimate of a desired statistic or feature from a databank or a collection of databanks. Disclosure can be achieved through one query or a series of queries to one or more databanks. With DNA information, queries might be directed at attaining numerical estimates of values or at deducing the state of an attribute of a person through a series of Boolean (yes-no) queries to multiple distributed databanks.

Several private laboratories already offer a DNA-banking service (sample storage in freezers) to physicians, genetic counselors, and, in some cases, anyone who pays for the service. Typically, such information as name, address, birth date, diagnosis, family history, physician's name and address, and genetic counselor's name and address is stored with the samples. That information is useful for local, independent bookkeeping and record management. But it is also ripe for statistical or correlative disclosure. Just the existence of a sample from a person in a databank might be prejudicial to the person, independently of any DNA related information. In some laboratories, the donor cannot legally prevent outsiders' access to the samples, but can request its withdrawal. A request for withdrawal might take a month or more to process. In most cases, only physicians with signed permission of the donor have access to samples, but typically no safeguards are taken to verify individual requests independently. That is not to say that the laboratories intend to violate donors' rights; they are simply offering a service for which there is a recognized market and attempting to provide services as well as they can. Much has been written on statistical databank systems and associated security issues.[3]

Guidelines for release of DNA samples and disclosure of DNA typing information must be designed to safeguard the rights of persons who, for one reason or another, get involved in a DNA typing (see Chapter 7 for further discussion) without burdening law-enforcement agencies and civil investigative authorities with unnecessarily protective policies.

The need for safeguards of DNA information has not been completely lost on lawmakers considering databank legislation. Some state legislation has addressed the issue. For example, the Virginia law[4] establishing a DNA profile databank for convicted offenders states that

> any person who, without authority, disseminates information contained in the databank shall be guilty of a Class 3 misdemeanor. Any person who disseminates, receives or otherwise uses or attempts to so use information in the databank, knowing that such use is for a purpose other than as authorized by law, shall be guilty of a Class 1 misdemeanor. Except as authorized by law, any person who, for purposes of having DNA analysis performed, obtains or attempts to obtain any sample submitted to the Division of Forensic Science for analysis shall be guilty of a Class 5 felony.

That passage reflects recognition of the potential for abuse of information derived from a sample (and of the sample itself) and incorporates sanctions to preclude it. In the first legal test[5] of the establishment of such databanks on convicted felons, Chief U. S. District Judge James C. Turk upheld the Virginia databank statute, offering the following opinion in regard to the issue of privacy:

> The stored information is available only to law enforcement personnel in furtherance of an official investigation of a criminal offense. Va. Code Ann. Section 19.2-310.6 (1990). In addition, the identifying information is disseminated to the law enforcement officer only if the sample provided by the officer matches a sample in the databank. Id. The procedures followed

are sufficiently stringent such that no person, including a law-enforcement official, may conduct random searches in the databank.

Although that is a good start, state laws should state explicitly the types of uses that can be authorized. In particular, in addition to the points made in the opinion just quoted, investigation of DNA samples or stored information for the purpose of obtaining medical information or discerning other traits should be prohibited, and violations should be punishable by law. Several states incorporate some of those specific protections into their statutes establishing DNA profile databanks. However, the committee urges all states to be systematic in defining authorized uses of information in DNA databanks.

METHODOLOGICAL STANDARDIZATION

Because of the incompatibility between DNA typing methods, federal, state, and local laboratories that wish to use a national DNA profile databank must all adopt a single standardized method for analyzing samples—both databank specimens and evidence specimens. Accordingly, the development of a national DNA databank has the potential advantage of acting as a driving force for standardization in forensic DNA typing, but the potential disadvantage of ossifying a rapidly moving technology.

It is broadly agreed that current RFLP typing methods constitute simply an initial approach that will be replaced in the next few years by procedures that are much easier to automate, much less expensive, and more informative. Premature development of a national databank based on current RFLP typing methods runs the risk of perpetuating a "dinosaur" technology in the face of better techniques.

The committee believes that it is too early to launch a comprehensive national DNA profile databank. However, it is appropriate to carry out pilot programs based on RFLP technology with the FBI and states that have active DNA typing efforts. The initial efforts should help to define the problems and issues that will be encountered in the fashioning of a comprehensive program. Such projects should be explicitly viewed as preliminary, with the clear expectation that the databank will be supplanted in the next several years by better methods.

Before even pilot projects can be begun, the degree of interlaboratory reproducibility—which is essential to the success of a databank—should be thoroughly documented. So far, there have been only a few interlaboratory-reproducibility studies to compare the ability of different laboratories to measure the same DNAs accurately under different circumstances. The National Institute of Standards and Technology (NIST), in concert with the Federal Bureau of Investigation (FBI) and the Technical Working Group on

DNA Analysis Methods (TWGDAM), sent samples to 22 laboratories in October 1990 (Dennis Reeder, personal communication, 1991); 12 laboratories have reported so far. The greatest differences were reported to be slightly less than 5%. The preliminary results are encouraging, but need to be followed by more extensive reproducibility testing before the efficacy of a national network based on this method can be demonstrated. Moreover, the committee urges that laboratories participating in any national databank be required to participate in continuing proficiency and reproducibility studies (by carrying out blind measurements of samples sent from a common source), to ensure that reproducibility does not drift over time.

COST VERSUS BENEFIT

An analysis of the costs and benefits of establishing DNA databanks is problematic at best. Costs will depend on a number of variables, such as methods, numbers of loci used, and types and numbers of samples to be tested. Benefits will depend on the populations included in the databank and the likelihood of finding matches. Moreover, costs and benefits must be reckoned in both monetary and nonmonetary terms.

Nonmonetary costs can include the risk of loss of privacy and the misuse and abuse of genetic information. Nonmonetary benefits can include prevention of future crimes. Those diverse elements cannot be weighed except in the context of societal values.

Concerning monetary costs, it is helpful to recall the comparison between latent fingerprints and DNA profiles. Collection of fingerprints from identified persons is inexpensive and relatively easily accomplished by persons with minimal technical training and background. Samples cost perhaps a few dollars; the cost reflects the personnel time involved in taking and filing the fingerprints. Although sample collection is simple, fingerprint databanks require sophisticated and expensive computer hardware and software. A typical state automated fingerprint identification system can cost $10 million. In contrast, DNA typing is time-consuming, is expensive, and requires extensive education, training, and quality-assurance measures. With current RFLP methods, blood must be obtained by venipuncture at an estimated cost of $20/sample. Storage methods and costs depend on the number of samples and the form in which they are preserved (liquid or dried blood, extracted DNA pellet, buffy coat, etc.). In any case, freezers, cryotubes, and labor can cost another $20/sample for storage. The cost of RFLP analysis can be estimated from fees charged by private laboratories: about $100-150/sample.[6] Thus, a single DNA profile can cost about $120-170, and constructing 10,000 DNA profiles could cost $1.2-1.7 million. However, DNA typing databanks do not require highly sophisticated or expensive computer hardware and software.

In short, ordinary fingerprints and DNA profiles have opposite economic characteristics. Ordinary fingerprint databanks have low variable costs and high fixed costs, and DNA typing databanks have high variable costs and comparatively low fixed costs. Those considerations imply that different decisions could be appropriate as to whether, when, and how to develop each kind of databank. For example, because of the high variable cost per sample, considerable thought must given to whose DNA profiles should be stored. To maximize the "return per sample," one should concentrate on persons convicted of crimes with documented high rates of recidivism, such as rape, as discussed below.

Cost analysis is made more difficult by the rapidity of change in DNA typing technology. For example, PCR-based methods might greatly reduce DNA typing costs: blood samples might be replaced with simple buccal swabs (i.e., cheek scraping); Southern blots might be replaced with non-gel-based formats; complicated scoring of the problematic continuous allele system used in RFLP analysis might be replaced with discrete mechanical allele scoring. Accordingly, today's cost assessments must be viewed as tentative.

WHOSE SAMPLES SHOULD BE INCLUDED?

In deciding whom to include in a DNA profile databank, it is necessary to consider the likely forensic utility of the data and the protection of individual privacy. It is helpful to consider six categories of people.

Samples from Convicted Offenders

DNA profile databanks containing profiles of criminal offenders must be justified on the basis of the likelihood of recidivism. The Bureau of Justice Statistics[7] found that, of the 108,580 persons released from prisons in 11 states in 1983, an estimated 63% were rearrested for a felony or serious misdemeanor within 3 years, 47% were reconvicted, and 41% returned to prison or jail (Table 5-1). They were charged with a total of 326,746 new offenses in the 3-year period; more than 50,000 charges were related to violent offenses, including approximately 2,000 homicides, 1,500 kidnappings, 1,300 rapes, 2,600 other sexual assaults, 17,000 robberies, and 22,600 other assaults. Of the prisoners who had been incarcerated for violent offenses, 60% were rearrested within 3 years for similar offenses. Recidivism rates were highest in the first year. Four of every 10 released prisoners were rearrested in the first year; nearly one-fourth were convicted of new crimes; and nearly one-fifth were returned to prison or sent to jail. Most rearrests occurred in the states in which the prisoners were released, although about 15% occurred in other states. Of course, high recidivism

TABLE 5-1 Recidivism Rates of Prisoners Released in 11 States in 1983, by Most Serious Offense[a]

	Fraction of Prisoners, %, Who Within 3 Years Were:			
Offense	Fraction of Prisoners, %	Rearrested	Reconvicted	Reincarcerated
All offenses	100.0	62.5	46.8	41.4
Violent offenses	34.6	59.6	41.9	36.5
Murder	3.1	42.1	25.2	20.8
Negligent manslaughter	1.4	42.5	27.9	21.8
Kidnapping	.6	54.5	35.7	31.3
Rape	2.1	51.5	36.4	32.3
Other sexual assault	2.1	47.9	32.6	24.4
Robbery	18.7	66.0	48.3	43.2
Assault	6.4	60.2	40.4	33.7
Other	.4	50.1	33.2	31.4
Property offenses	48.3	68.1	53.0	47.7
Burglary	25.8	69.6	54.6	49.4
Larceny and theft	11.2	67.3	52.2	46.3
Motor vehicle theft	2.6	78.4	59.1	51.8
Arson	.7	55.3	38.5	32.3
Fraud	5.5	60.9	47.1	43.3
Stolen property	1.7	67.9	54.9	50.5
Other	.8	54.1	37.3	33.9
Drug offenses	9.5	50.4	35.3	30.3
Possession	1.2	62.8	40.2	36.7
Trafficking	4.5	51.5	34.5	29.4
Other and unspecified	3.9	45.3	34.5	29.1
Public-order offenses	6.4	54.6	41.5	34.7
Weapons	2.2	63.5	46.7	38.1
Other	4.2	49.9	38.9	33.0
Other offenses	1.1	76.8	62.9	59.2

[a]Data from Beck and Shipley.[7]

rates alone do not demonstrate the utility of a databank of DNA profiles of convicted offenders. One must also ask: What fraction of crimes committed by repeat offenders do not themselves lead to rearrest and reconviction? What fraction would end in rearrest and reconviction if a DNA profile databank were available?

The first question obviously is impossible to answer explicitly. However, the FBI's *Uniform Crime Reports* states that there are about 20,000 murders and 100,000 forcible-rape cases per year. It is estimated that 30% of all murder cases and 70% of all rape cases are never closed by arrest (John Hicks, personal communication, 1990). It should also be pointed out that only an estimated 50% of rapes are in fact even reported.

The second question is also difficult to answer, but it is clear that crimes of most types will not afford the opportunity to recover relevant biological evidence that will allow the police to identify an unknown suspect—i.e., the perpetrator's own body fluids. They include larcenies, burglaries, and assaults, for which ordinary fingerprints are frequently found. The major exception is rape, for which semen samples can be recovered in many cases and might provide prima facie evidence of sexual intercourse. In a small minority of homicides, blood, hair, or tissue samples from the perpetrator are left at the scene of the crime (e.g., because of a fight at the scene).

A DNA profile databank would thus be valuable primarily in investigating forcible rape, although the databank would be useful for some other investigations. State legislatures considering setting up such databanks should weigh the benefits in terms of solved rape cases and the costs in terms of collecting samples from persons likely to commit rapes (primarily, it seems, convicted sex offenders). Initial state efforts to develop DNA profile databanks were indeed aimed at sex offenders. Interestingly, some states rapidly expanded their programs to include all convicted offenders—without explicit weighing of the potential benefits of possessing such persons' patterns for solving crimes and the potential costs.

The above discussion justifies the development of a databank of DNA profiles of unknown subjects (open cases) and of offenders convicted of violent sex crimes. Such a databank would provide law enforcement with a powerful tool in linking sexual-assault cases through DNA profiles and tracking the activities of serial rapists. In light of recidivism and the continuing increase in reported rapes in this country, a databank of convicted sex offenders would provide investigators with a logical first place to look for assistance in solving unknown-offender sexual-assault cases.

Samples from Suspects

DNA typing profiles of suspects might also be useful in associating a person with open or unsolved cases pending in other jurisdictions or states. Although a suspect's DNA profile might ultimately be entered into a convicted-felon databank, there would no doubt be a substantial period during which a suspect might engage in other criminal activities. Thus, in the case of a serial rapist, a person under suspicion and investigation for one offense, might be responsible for several later offenses for which he is not suspected. Therefore, if a DNA profile of a suspect is entered into a databank, it would be available to be searched against future unsolved cases.

Samples from Victims

To protect their privacy, victims' DNA profiles should never be entered into a national databank or searched against such a databank, with the possible exception of cases of abduction, in which it might be desirable for the victim's information to be stored and accessible to law-enforcement officials. In any exceptional case, prior permission of the victim, the victim's legal guardian, or a court should be required, and the victim's DNA should be removed from the databank when it can no longer serve the purpose for which it was entered.

Samples from Missing Persons and Unidentified Bodies

This portion of the databank would contain DNA profiles from unidentified bodies, body parts, and bone fragments. These would provide the greatest benefit when DNA profiles from immediate relatives (parents) could be used to reconstruct the DNA profile of a missing person for comparison. Although there would be immediate benefits from the development of these types of data, the actual number of relevant cases would be small, compared with the number of sexual assaults by unknown persons.

Crime-Scene Samples from Unidentified Persons

DNA profile evidence found at the scene of a crime should be stored and accessible to legally authorized investigators. Such samples might be useful for recognizing serial or multiple crimes even before a perpetrator is found and will be equally useful once a perpetrator has been identified. It might be useful to have additional cross-referenced information accessible at the national level, including modus operandi or other attributes for correlation as part of an investigation.

Samples from Members of the General Population

Some observers have suggested that a DNA profile databank should not be limited to criminals, but should aim, at least in the long term, to store DNA profiles from the entire general public. It is argued that many groups in the general public are already required to be fingerprinted for various security and identification purposes and the same justification could be applied to DNA profiles; furthermore, if the databanks contained everyone, rather than just previous offenders, the chance of identifying perpetrators would be much greater.

The committee does not find those arguments persuasive. For identification and security purposes, DNA profiles would add nothing to ordinary fingerprints, because ordinary fingerprints already provide a complete identifier and are far more likely to be recovered in connection with security breaches than are blood samples that are amenable to DNA analysis. As for identifying perpetrators, there is no doubt that the system would have some effect. However, Americans have generally been reluctant to allow the creation of national identification systems, and DNA profiling poses a special risk of invasion of privacy (concerning personal and medical traits). We caution against moving in that direction. Finally, we note that current technology is far too expensive to contemplate the creation of such a large databank.

Samples from Anonymous Persons for Population Genetics

The committee notes that statistical databanks of random population samples are required for estimating allele frequencies, as described in Chapter 3. To protect the privacy of persons whose only role is to make up a statistical sample, their identities should never be retained in a databank, and the databanks should never be searched for matches in connection with investigations.

SAMPLE STORAGE

Another difficult issue is the storage and maintenance of DNA samples themselves (or any reusable products of the typing process), as opposed to DNA profiles. In principle, retention of DNA samples creates an opportunity for misuse—i.e., for later testing to determine personal information. In general, the committee discourages the retention of DNA samples.

However, there is a practical reason to retain DNA samples for short periods. Because DNA technology is changing so rapidly, we expect the profiles produced with today's methods to be incompatible with tomorrow's methods. Accordingly, today's profiles will need to be discarded and replaced with profiles based on the successor methods. It would be extremely expensive and inefficient to have to redraw blood samples for retyping. We are therefore persuaded that retention of samples after typing should be permitted for the short term—only during the startup phase of DNA profile databanks. As databanks become established and technology stabilizes somewhat, samples should be destroyed promptly after typing.

INFORMATION TO BE INCLUDED AND MAINTAINED IN A DATABANK

It is worth commenting on the nature of the information that should be stored in a DNA profile databank.

• Submitting-agency information should include the location of the agency, its telephone number, names of the analysts who conducted the DNA typing, the name of the person who entered the data into the databank, and agency contact information.

• Sample information should include entries that describe the type of sample (body-fluid stain, tissue, or known blood sample) and a unique sample identifier, the condition of the sample, unusual handling and storage, and other factors that might affect the quality of the DNA and the evaluation of partial patterns.

• The DNA type at a locus must be entered in standard nomenclature. For example, for RFLP typing, fragment-size data from each locus successfully probed should be entered as the number of basepairs determined for each fragment. Sizing data for the human-DNA control should also be entered.

• Entries into the convicted-offender files should include the name of the offender, dates of offenses and convictions, and DNA profile data. Only the profile index should be centrally stored. Case data should be stored locally, and their distribution should be under the control of the local agency.

RULES ON ACCESSIBILITY

Computer security should be ensured through use of the best available practices and technologies. Access to the databank should be limited to a small number of legally authorized persons and should be limited to what is required for specific official investigations. All instances of access should be audited and archived. An excellent discussion of computerized audit-trail systems is available.[8]

If the computer system and associated databank are to be made available for remote access by cooperating state and federal agencies, such as by telephone or networked by other means, the access mechanism (i.e., the network switch) should be made available only for specific, authorized remote-access sessions; that is, the system should not be continuously available to remote users. This type of limited access can be achieved either administratively or physically; it is a simple and inexpensive means of safeguarding sensitive information and is common practice in many national security situations. For example, secure computers are virtually never connected to unsecured computers at national defense laboratories; when newspaper headlines make statements that computers at these facilities have been breached, it has been the case that the computers were unsecured and not connected to the secure computers. In many cases, these unsecured computers have telecommunication connections available to employees for routine use, but they do not contain security information.

STATISTICAL INTERPRETATION OF DATABANK MATCHES

The distinction between finding a match between an evidence sample and a suspect sample and finding a match between an evidence sample and one of many entries in a DNA profile databank is important. The chance of finding a match in the second case is considerably higher, because one does not start with a single hypothesis to test (i.e., that the evidence was left by a particular suspect), but instead fishes through the databank, trying out many hypotheses.

If a pattern has a frequency of 1 in 10,000, there would still be a considerable probability (about 10%) of seeing it by chance in a databank of 1,000 people. Although there are statistical methods for correcting for such multiple testing, the committee considers that approach unwise, because it requires that the population frequency estimates of genotypes are accurate to a degree that is unlikely to be achieved (because sample sizes are limited). There is a far better solution: When a match is obtained between an evidence sample and a databank entry, the match should be confirmed by testing with additional loci. The initial match should be used as probable cause to obtain a blood sample from the suspect, but only the statistical frequency associated with the additional loci should be presented at trial (to prevent the selection bias that is inherent in searching a databank). Forensic DNA typing laboratories should recognize that they will require additional loci beyond those used in the databank to prove a case against a suspect. Preparations should be begun now to have additional loci characterized and available for general use before any DNA profile databank comes into common use.

STATUS OF DATABANK DEVELOPMENT

There have already been state and federal efforts toward the creation of DNA profile databanks. We review them briefly here.

State Level

According to a recent FBI survey,[9] 27% of 177 forensic science laboratories responding indicated that they have legislative authority or a mandate to construct a databank for their own jurisdictions to match DNA profiles. An additional 38% believed that such authority or mandate was likely by 1991. According to an Office of Technology Assessment survey conducted in late 1989,[10] at least 17 states had passed or were considering legislation creating statewide DNA databanks. The persons to be included in the databanks range from sex offenders to all felons. Since the time of that survey, the number has no doubt increased. Therefore, it is obvious that many state legislatures recognize the potential benefit of a DNA databank as an important investigative tool and that such databanks will become a reality. Many states are already collecting samples in earnest, although at this writing no databanks are operative.

Federal Level

The FBI and TWGDAM have proposed the creation of a national DNA profile databank system, including one statistical and three investigative databanks. The statistical databank would include DNA profiles of randomly selected unrelated persons and would be built collaboratively and maintained by the FBI for use by all forensic laboratories. The investigative databanks would contain DNA profiles of body fluids from the scenes of crimes for which suspects have been identified, convicted offenders, and bodies, body parts, and bone fragments of unidentified persons. In the proposed national DNA profile databank system, individual law-enforcement agencies (forensic laboratories) would contribute DNA profiles (without personal information) to a centralized databank, but retain absolute control of their own case records. The national databanks would reference the sources of the profiles, but case data would be secured and controlled by the state and local agencies.

In the national program, the FBI would play the lead role. It would coordinate quality assurance with a technical advisory group to implement appropriate guidelines; coordinate with other agencies that have a law-enforcement interest in the development of the databank; provide hardware and software for the databank server and for state access to the databank; provide hardware to store and back up the databank server; provide training for states in forensic DNA technology, quality control, and databank access; determine formats for databank input and output; update index with new state and federal submissions; assemble population data for all probes used and calculate and disseminate population frequencies; and modify the system to accommodate new DNA typing methods.

State and local agencies would be responsible for performing DNA analyses of samples with consensus methods; submitting new information in a specified format for incorporation into the databanks; guaranteeing the quality of their new submissions; providing hardware and software for state image-analysis workstations for telephone access to centralized index; maintaining centrally indexed case files for as long as they remain in the index; and providing relevant information from case files that are indexed centrally to other law-enforcement agencies, which subscribe when requested.

Just as the Department of Defense keeps dental records and fingerprints (with the FBI) of American soldiers, it is seeking funding to collect blood samples from each soldier and establish a DNA profile databank. When a soldier is killed and cannot be identified with usual methods, a sample of tissue, blood, or bone marrow from the remains would be subjected to DNA analysis for comparison with entries in the databank. There are 3.3 million active and reserve members of the armed forces. Given the costs associated with the current technology, a DNA databank of such scope would not be amenable to RFLP analysis. The Armed Forces Institute of Pathology therefore proposes to begin collecting and storing samples while working on the development of a DNA analysis method, which when perfected will be much less expensive and time-consuming than existing RFLP methods.

A databank of military personnel could also offer ancillary forensic applications: criminal investigations conducted by criminal investigation divisions of the armed forces could be aided in the same manner as those of other law-enforcement agencies, identification of subjects for security purposes could be enhanced, and identification of urine samples from disputed sources for drug testing could be verified.

The present committee has not been asked to comment on this program; we simply acknowledge its existence.

MODEL COOPERATIVE INFORMATION RESOURCE

Local autonomy as to databank structure and function is recommended, for several reasons: a databank can be tailored to meet local needs, the local databank administrator will not have to rely on outside entities for maintenance and change, and security can best be managed with smaller, discrete, well-understood databanks. That is not to say that standards and guidelines should be avoided. On the contrary, very strict regulations, standards, and guidelines for all aspects of the operation should be enforced and monitored. Databank requirements involve determining what a system must accomplish; there are typically many alternative implementation details that can accomplish the same goals.

The experimental protocols used to derive DNA profiles will probably continue to change as the associated technologies continue to mature. That presents a problem that is common in databank applications when the underlying science is in flux: maintaining data integrity while keeping the system current with the most appropriate technology. It will be challenging, but necessary to ensure competence. In practice, that means designing for change, which requires partitioning the problem into two domains—one that is relatively stable and one that is relatively dynamic. For example, data within the sample context are relatively stable, whereas those associated with experiments and derived data are relatively dynamic.

Figure 5-1 is a high-level data flow diagram that shows one possible model for the flow of information from state or regional laboratories to a

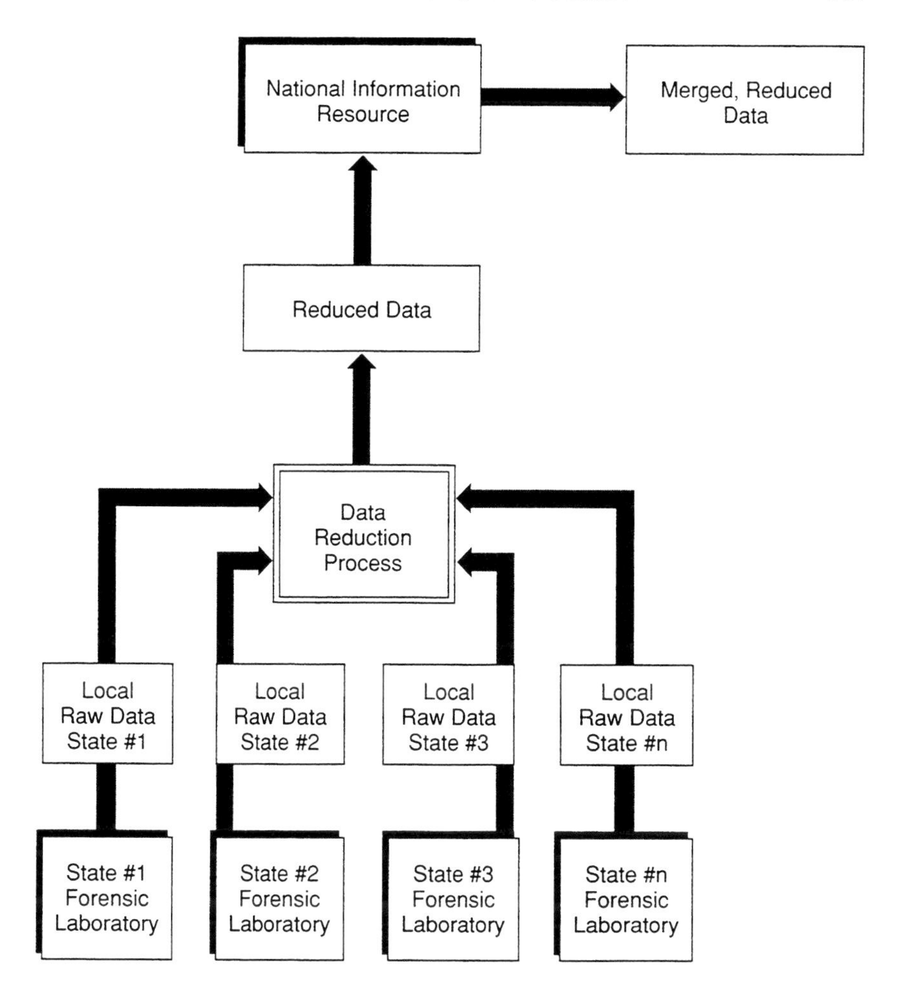

FIGURE 5-1 Hypothetical national information resource. Data flow starts with forensic laboratories in various states that provide raw data. Data reduction process provides information to national information resource databank. Merged and reduced data are provided only to authorized users.

national information resource. Each state or region could have several participating local facilities, but for simplification it is recommended that each state or region have one official clearinghouse for locally derived information. Regional facilities could range from situations where a state has several laboratories serving a high-volume workload, as in California, to a regional group of states that each have only periodic and low-volume workloads. All the locally generated raw data and results would be stored at the state or regional level. Thus, all the information concerning the sample, experimental, and result contexts would be stored at the state or regional level; only data associated with the result context would be accessed at the national level. The box labeled "Data Reduction Process" in the center of the figure ideally represents a standardized method for DNA typing that all laboratories use.

The hypothetical national information resource shown in Figure 5-1 does not necessarily represent a physical entity, but could be simply a view of derived data from all the various regional databanks. A view would be achieved by having access software sitting "on top" of the various state or regional databanks. The software would have distinctly different requirements for level of access to data in the databank. For example, "outside views" would only need access to VNTR profiles and some arbitrary identification number; no further information at this first level of access would be required for initial identification searching.

SUMMARY OF RECOMMENDATIONS

• In principle, a national DNA profile databank should be created that contains information on felons convicted of violent crimes with high rates of recidivism. The case is strongest for felons who have committed rape, because perpetrators typically leave biological evidence (semen) that could allow them to be identified. The case is somewhat weaker for violent offenders who are most likely to commit homicide as a recidivist offense, because killers leave biological evidence only in a minority of cases. The wisdom of including other offenders depends primarily on the rate at which they are likely to commit rape, because rape is the crime for which the databank will be of primary use.

There are a number of scenarios that illustrate the point that the databank need not be limited to persons convicted of specified crimes.

• The databank should also contain DNA profiles of samples from unidentified persons collected at the scenes of violent crimes.

• Databanks containing DNA profiles of members of the general population (as exist for ordinary fingerprints for identification purposes) are not appropriate, for reasons of both privacy and economics.

• DNA profile databanks should be accessible only to legally authorized persons and should be stored in a secure information resource.

• Legal policy concerning access and use of both DNA samples and DNA databank information should be established before widespread proliferation of samples and information repositories. Interim protection and sanctions against misuse and abuse of information derived from DNA typing should be established immediately. Policies should explicitly define authorized uses and should provide for criminal penalties for abuses.

• Although the committee endorses the concept of a limited national DNA profile databank, we doubt that existing RFLP-based technology provides a wise long-term foundation for such a databank. We expect current methods to be replaced soon with techniques that are simpler, easier to automate, and less expensive—but incompatible with existing DNA profiles. Accordingly, we do not recommend establishing a comprehensive DNA profile databank yet.

• For the short term, we recommend the establishment of pilot projects that involve prototype databanks based on RFLP technology and consisting primarily of profiles of violent sex offenders. Such pilot projects could be worthwhile for identifying problems and issues in the creation of databanks. However, in the intermediate term, more efficient methods will replace the current one, and the forensic community should not allow itself to become locked into an outdated method.

• State and federal laboratories, which have a long tradition and much experience with the management of other types of basic evidence, should be given primary responsibility, authority, and additional resources to handle forensic DNA testing and all the associated sample-handling and data-handling requirements.

• Private-sector firms should not be discouraged from continuing to prepare and analyze DNA samples for specific cases or for databank samples, but they must be held accountable for misuse and abuse to the same extent as government-funded laboratories and government authorities.

• Discovery of a match between an evidence sample and a databank entry should be used only as the basis for further testing using markers at additional loci. The initial match should be used as probable cause to obtain a blood sample from the suspect, but only the statistical frequency associated with the additional loci should be presented at trial.

REFERENCES

1. Wilson T. Automated fingerprint identification systems. Law Enforc Technol. 1986. Aug-Sep:17-20, 45-48.
2. Dalenius T. Towards a methodology and statistical disclosure control. Statistik Tidskrift. 15:213-225, 1977.
3. Adam NR, Wortmann JC. Security-control methods for statistical databases: a comparative study. ACM Comput Surv. 21:515-556, 1989.
4. Code of Virginia, Title 19.2-310.6. Unauthorized uses of DNA data bank; forensic samples; penalties (1990, c. 669).
5. *Jones v. Murray*, 763 F. Supp. 842 (W.D. Va. 1990).
6. Cellmark Diagnostics. Proposal for DNA databasing, Division of Criminal Investigation Forensic Laboratory, South Dakota. Dec. 21, 1990.

7. Beck A, Shipley B. Recidivism of prisoners released in 1983. Bureau of Justice Statistics Special Report NCJ-116261. Washington, D.C., 1989.
8. Lunt TF, Tamaru A, Gilham F, Jagannathan R, Neuman PG, Jalali C. IDES: a progress report. Proceedings of the Sixth Annual Computer Security Applications, Tucson, Arizona: ACM Press, 1990.
9. Miller J. The outlook for forensic DNA testing in the United States. Crime Lab Digest. 17(Suppl. 1):1-14, 1990.
10. U.S. Congress, Office of Technology Assessment. Genetic witness: forensic uses of DNA test. OTA-BA-438. Washington, D.C.: U.S. Government Printing Office, 1990.

6

Use of DNA Information in the Legal System

This chapter provides an overview of how DNA evidence might be used in the investigation and prosecution of crimes and in civil litigation. The DNA typing discussed in this chapter is mainly standard single-locus RFLP typing on Southern blots without apparent band shifting; i.e., it is the technique most often considered by the courts to date. We begin with a discussion of the investigation stage, but devote most of our attention to admissibility. In that context, we review some of the rapidly growing number of cases involving admissibility. Discussion of case law is intended mainly to highlight specific issues and is not intended to be comprehensive. Finally, we make a series of practical recommendations, with judges especially in mind.

To produce biological evidence that is admissible in court in criminal cases, forensic investigators must be well trained in the collection and handling of biological samples for DNA analysis. They should take care to minimize the risk of contamination and ensure that possible sources of DNA are well preserved and properly identified. As in any forensic work, they must attend to the essentials of preserving specimens, labeling, and the chain of custody and to any constitutional or statutory requirements that regulate the collection and handling of samples. The Fourth Amendment provides much of the legal framework for the gathering of DNA samples from suspects or private places, and court orders are sometimes needed in this connection.

Wherever possible, a preserved sample should be large enough to enable the defense to obtain an independent RFLP analysis, but there should almost always be enough at least for PCR analysis, a technique likely to be widely used in forensics in the near future for amplification of the DNA in the evidentiary sample. All materials relied on by prosecution experts must be available to defense experts, and vice versa. The laboratories used for analysis must be reliable and should be willing to meet recognized standards of disclosure.

In civil (noncriminal) cases—such as paternity, custody, and proof-of-death cases—the standards for admissibility must also be high, because DNA evidence might be dispositive. The relevant federal rules (403, 702-706) and most state rules of evidence do not distinguish between civil and criminal cases in determining the admissibility of scientific data. In a civil case, however, if the results of a DNA analysis are not conclusive, it will usually be possible to obtain new samples for study. As in criminal cases, laboratories and other interested parties must treat evidence according to established protocols.

The advent of DNA typing technology raises two key issues for judges: determining *admissibility* and explaining to jurors the appropriate standards for *weighing* evidence. A host of subsidiary questions with respect to how expert evidence should be handled before and during a trial to ensure prompt and effective adjudication apply to all evidence and all experts and are not dealt with in this chapter.

ADMISSIBILITY

In the United States, there are two main tests for admissibility of scientific information from experts. One is the Frye test, enunciated in *Frye v. United States*.[1] The other is a "helpfulness" standard found in the Federal Rules of Evidence and many of its state counterparts. In addition, several states have recently enacted laws that essentially mandate the admission of DNA typing evidence.

The Frye Test

The test for the admissibility of novel scientific evidence enunciated in *Frye v. United States* has been the most frequently invoked one in American case law. A majority of states profess adherence to the *Frye* rule, although a growing number have adopted variations on the helpfulness standard suggested by the Federal Rules of Evidence.

Frye predicates the admissibility of novel scientific evidence on its general acceptance in a particular scientific field: "While courts will go a long way in admitting expert testimony deduced from a well-recognized scientific principle or discovery, the thing from which the deduction is made must be sufficiently established to have gained general acceptance in the particular field in which it belongs."[2] Thus, admissibility depends on the quality of the science underlying the evidence, as determined by scientists themselves. Theoretically, the court's role in this preliminary determination is quite limited: it should conduct a hearing to determine whether the scientific theory underlying the evidence is generally accepted in the relevant scientific community and to determine that the specific techniques used are reliable for their intended purpose.

In practice, the court is much more involved. The court must determine which scientific fields experts should be drawn from. Complexities arise with DNA typing, because the full typing process rests on theories and findings that pertain to various scientific fields. For example, the underlying theory of detecting polymorphisms is accepted by human geneticists and molecular biologists, but population geneticists and statisticians might differ as to the appropriate method for determining the population frequency of a genotype in the general population or in a particular geographic, ethnic, or other group. The courts often let experts on a process, such as DNA typing, testify to the various scientific theories and assumptions on which the process rests, even though the experts' knowledge of some of the underlying theories is likely to be at best that of a generalist, rather than a specialist.

When a process is new and complex, a court should recognize that the expertise of more than one discipline might be necessary to explain it. That is the case when the admissibility of DNA evidence is judged as a matter of first impression. Among the issues raised is the validity of the assumptions that (1) except for identical twins, each person's DNA is unique, (2) the technique used allows one to determine whether two DNA samples show the same patterns at particular loci, and (3) the statistical methods used and the available population databanks allow one to assess the probability that two DNA samples from different persons would by chance have the same patterns at the loci studied. Even if those assumptions are accepted, there is the important question of whether (4) the laboratory's procedures and analyses in the case in question were performed in accordance with accepted standards and provide reliable estimates of the probability of a match.

Assumption 1—that, with the exception of identical twins, each person's DNA is unique—is so well established in human molecular genetics that a court is justified in judicially noticing it, even in the context of a *Frye* hearing.

Assumption 2 concerns the validity of procedures for extracting DNA from samples of blood, semen, and other materials and analyzing it for the presence and size of polymorphisms. With regard to application in scientific research, the validity is sufficiently well established in the case of RFLP analysis with Southern blots that judicial notice is also appropriate. With regard to the application in forensic science, however, additional questions of reliability are raised. For example, forensic DNA analysis frequently involves the use of small, possibly contaminated samples of unknown origin, such as a dried blood stain on a piece of clothing. Some experts have questioned the reliability of DNA analysis of samples subjected to "crime scene" conditions. In addition (as noted in Chapters 2 and 3), the details of the particular techniques used to perform DNA typing and to resolve ambiguities evoke a host of methodological questions. It is usually appropriate to evaluate these matters case by case in accordance with the standards and cautions contained in earlier portions of this report, rather than generally excluding DNA evidence. Of particular importance once such a system of quality assurance is established would be a demonstration that the involved laboratory is appropriately accredited and its personnel certified. Some aspects (such as the validity of the theory underlying RFLP analysis) might be so well established that judicial notice is warranted. Others (such as quantitative correction of band shifting with a single monomorphic fragment) might not be sufficiently well established to justify admissibility.

Assumption 3—related to the adequacy of statistical databanks used to calculate match probabilities—rests on unproven foundations. Many experts question the adequacy of current databanks for making probability estimates, and the use of multiplicative modes of combining probabilities are also open to serious question (see Chapter 3). The solution, however, is not to bar DNA evidence, but to ensure that estimates of the probability that a match between a person's DNA and evidence DNA could occur by chance are appropriately conservative (as described in Chapter 3).

The validity of assumption 4—that the analytical work done for a particular trial comports with proper procedure—can be resolved only case by case and is always open to question, even if the general reliability of DNA typing is fully accepted in the scientific community. The DNA evidence should not be admissible if the proper procedures were not followed. Moreover, even if a court finds DNA evidence admissible because proper procedures were followed, the probative force of the evidence will depend on the quality of the laboratory work. More control can be exercised by the court in deciding whether the general practices in the laboratory or the theories that a laboratory uses accord with acceptable scientific standards. Even if the general scientific principles and techniques are accepted by experts in the field, the same experts could testify that the work done in a particular case was so flawed that the court should decide that, under *Frye*, the jury should not hear the evidence.

The *Frye* test sometimes prevents scientific evidence from being presented to a jury unless it has sufficient history to be accepted by some subspecialty of science. Under *Frye*, potentially helpful evidence may be excluded until consensus has developed.[3] By 1991, DNA evidence had been considered in hundreds of *Frye* hearings involving felony prosecutions in more than 40 states. The overwhelming majority of trial courts ruled that such evidence was admissible; there have been some important exceptions, however.

The first scientifically thorough *Frye* hearing concerning DNA typing was conducted in *People v. Castro*,[4] in which a New York trial court concluded that the theory underlying DNA typing is generally accepted by scientists in genetics and related fields, that forensic DNA typing has also been accepted and is reliable, but that the technique as applied in the particular case was so flawed that evidence of a match was inadmissible (although evidence of an exclusion was admissible). The *Castro* court stated that the focus of the *Frye* test as applied to DNA typing (or any other novel scientific evidence of similar complexity) must include its application to the particular case. It held that flaws in the application are not simply questions as to the weight to be given the evidence by the jury, but go to admissibility as determined by the judge.[5] *Castro* determined that there were serious flaws in the laboratory's declaration of a match between two samples, for a number of reasons, including the presence of several anomalous bands. The court did not credit the laboratory's explanation of the reasons for the anomalies and criticized its failure to perform adequate follow-up testing. In addition, the court concluded that the laboratory's population-frequency databank could not provide an accurate estimate of the likelihood that the defendant was the source of the DNA. The court's analysis and findings were careful, and they have generally been approved by experts in the field.

In November 1989, the Supreme Court of Minnesota, deciding *State v. Schwartz*,[6] became the first appellate court to reject the use of DNA evidence analyzed by a forensic laboratory. In answering a certified question, the court noted that "DNA typing has gained general acceptance in the scientific community." Nevertheless, the court went on to hold that admissibility of specific test results in a particular case hinges on the laboratory's compliance with appropriate standards and controls and on the availability of its testing data and results. It held that, "because the laboratory in this case did not comport with these guidelines, the test results lack foundational adequacy and, without more, are thus inadmissible." One matter that troubled the court was the failure of the testing laboratory to reveal underlying population data and testing methods. The court noted that the reliability of a test implies that it could be subjected to an independent scientific assessment of the methods, including replication of the test. Because such independent assessment had not occurred and could not take place, owing to the laboratory's secrecy, the court held that the results were inadmissible. In addition, the court was concerned that the testing laboratory (1) had admitted having falsely identified two of 44 samples as coming from the sample subject during a proficiency test performed by the California Association of Crime Laboratory Directors and (2) had not satisfied relevant validation protocols used by the FBI. In that regard, *Schwartz* makes a good case for requiring laboratories to meet particular standards before they may provide analysis of evidence to juries. *Schwartz* also held that the use of population-frequency statistics must be limited, because "there is a real danger that the jury will use the evidence as a measure of the probability of the defendant's guilt or innocence, and the evidence will thereby undermine the presumption of innocence, erode the values served by the reasonable double standard, and dehumanize our system of justice."[7] The decision in *Schwartz* was influenced by Minnesota's unique position in limiting the use of probability estimates in trials.[8]

A new Minnesota statute not considered in *Schwartz* specifically requires judges to admit population-frequency data generated by DNA testing. Thus, it is not clear how influential *Schwartz* will be in its home state. Nevertheless, the Minnesota judges' skepticism about statistical analysis is shared by other judges. Particularly in regard to DNA typing, the manner in which probabilities should be calculated requires great care.

In *Cobey v. State*,[9] the Maryland Court of Special Appeals reached a conclusion opposite to *Schwartz*, holding that evidence of DNA analysis from the same laboratory that figured in *Schwartz* was admissible and finding that the laboratory's databank was sound. The *Cobey* court was impressed by the absence of expert testimony contradicting that in favor of admissibility. It did caution, however, that "we are not, at this juncture, holding that DNA fingerprinting is now admissible willy-nilly in all criminal trials." In 1989, Maryland became one of a growing number of states to enact a law recognizing the admissibility of DNA evidence.

Admissibility According to the Helpfulness Standard

The Federal Rules of Evidence, without specifically repudiating the *Frye* rule, adopt a more flexible approach. Rule 702 states that,

> if scientific, technical or other specialized knowledge will assist the trier of fact to understand the evidence or to determine a fact in issue, a witness qualified as an expert by knowledge, skill, experience, training, or education, may testify thereto in the form of an opinion or otherwise.

Rule 702 should be read with Rule 403, which requires the court to determine the admissibility of evidence by balancing its probative force against its potential for misapplication by the jury. In determining admissibility, the court should consider the soundness and reliability of the process or technique used in generating evidence; the possibility that admitting the evidence would overwhelm, confuse, or mislead the jury; and the proffered connection between the scientific research or test result to be presented and particular disputed factual issues in the case.[10]

The federal rule, as interpreted by some courts, encompasses *Frye* by making general acceptance of scientific principles by experts a factor, and in some cases a decisive factor, in determining probative force.[11] A court can also consider the qualifications of experts testifying about the new scientific principle,[12] the use to which the new technique has been put,[13] the technique's potential for error,[14] the existence of specialized literature discussing the technique, and its novelty.[15]

With the helpfulness approach, the court should also consider factors that might prejudice the jury. One of the most serious concerns about scientific evidence, novel or not, is that it possesses an aura of infallibility that could overwhelm a jury's critical faculties. The likelihood that the jury would abdicate its role as critical fact-finder is believed by some to be greater if the science underlying an expert's conclusion is beyond its intellectual grasp. The jury might feel compelled to accept or reject a conclusion absolutely or to ignore evidence altogether.[16] However, some experience indicates that jurors tend not to be overwhelmed by scientific proof and that they prefer experiential data based on traditional forms of evidence. Moreover, the presence of opposing experts might prevent a jury from being unduly impressed with one expert or the other. Conversely, the absence of an opposing expert might cause a jury to give too much weight to expert testimony, on the grounds that, if the science were truly controversial, it would have heard the opposing view. Other possible difficulties with the presentation of DNA expert evidence include the possibility of jury confusion and an inordinate consumption of trial time.[17] Nevertheless, if the scientific evidence is valid, the solution to those possible problems is not to exclude the evidence, but to ensure through instructions and testimony that the jury is equipped to consider rationally whatever evidence is presented.

In determining admissibility with the helpfulness approach, the court should consider a number of factors in addition to reliability. The first is the significance of the issue to which the evidence is directed. If the issue is tangential to the case, the court should be more reluctant to allow a time-consuming presentation of scientific evidence that might confuse the jury. Second, the availability and sufficiency of other evidence might make expert testimony about DNA superfluous. And third, the court should be mindful of the need to instruct and advise the jury to eliminate the risk of prejudice.[18]

Cases on Admissibility of DNA Evidence Under the Federal Rules

As with the *Frye* rule, courts applying the federal rules or conforming state rules must consider whether the particular techniques used in a particular case pass scientific muster. Three federal courts have now conducted thorough hearings on the admissibility of DNA evidence, with two courts finding it admissible and one ruling it inadmissible.

The U.S. District Court for the District of Vermont conducted a detailed analysis in *United States v. Jakobetz.*[19] It reviewed the literature and FBI practices. Despite a strong attack from the defense and its experts, the court found that the DNA evidence was "highly reliable" and that its probative value outweighed the potential for prejudice.[20] Strict application of the *Frye* test was rejected in accordance with Second Circuit standards.[21]

After a thorough hearing that focused on FBI protocols, the U.S. magistrate for the Southern District of Ohio in *United States v. Yee*[22] also wrote a detailed analysis with conclusions essentially tracking those in the Vermont case. (Interestingly, an Arizona trial court considering the admissibility of DNA typing in *State v. Despain*[23] carefully studied the transcript of *Yee*, but reached a conclusion opposite to it. That might have been because it also reviewed the transcript of another hearing in which four additional defense experts challenged FBI protocols. Finding that there was a legitimate scientific controversy as to the validity of DNA testing and that it had not gained general acceptance, the court in *Despain* refused to admit evidence analyzed by the FBI laboratory.)

Most recently, the Superior Court for the District of Columbia reached the opposite conclusion and held DNA typing inadmissible. In *U.S. v. Porter*[24], the court ruled that the technical reliability of DNA typing was generally accepted, but that the FBI's method for calculating the probability of a coincidental match was not. The court ruled that the scientific foundation of these probability calculations bears on the admissibility (and not simply the weight) of the evidence. Applying the *Frye* standard, the court found that "there is a controversy within the scientific community [on this issue] which has generated further study, the results of which will soon be available for scrutiny. It is after these studies and others . . . when the court should be called upon to admit DNA evidence."

In addition, a number of state courts that apply analogues of the federal rules have considered the admissibility of DNA evidence. In *Andrews v. State,*[25] a Florida court of appeals (the first higher-level state court to consider DNA evidence) determined that the relevance approach was applicable under the Florida evidence code that tracks the federal rules. The court admitted the evidence presented by the plaintiff's three scientific experts, two of whom worked for a private testing laboratory; the defense called no experts. The court concluded that the DNA typing evidence offered by the plaintiff was clearly helpful to the jury. With respect to the possibility of prejudice, the court found that DNA typing is not particularly "novel," in that it had been used in nonforensic applications for 10 years. The issue of differences between scientific applications and forensic applications were not raised by the defense. The court also noted the existence of specialized literature about the technique. As for the possibility of erroneous test results, the court credited testimony that an error in the testing process would mean that there would be no result, rather than a false-positive or false-negative result. The court also credited the efficacy of the laboratory's control runs and approved the use of statistical data to determine the probability of a match.

In *Spencer v. Commonwealth*,[26] the Supreme Court of Virginia affirmed a trial court's finding that evidence derived from RFLP analysis was sufficiently reliable to be admitted. The trial court heard testimony from three experts for the prosecution in molecular biology and genetics. The defense called no expert witnesses. The trial court credited testimony that there is no risk of false positives, that the testing techniques are reliable and generally accepted in the scientific community, and that the particular test was conducted in a reliable manner.

At a later proceeding involving the same defendant, the Supreme Court of Virginia held that evidence based on a sample analysis that used a PCR technology was admissible. In discussing the standard for admitting novel scientific evidence, it rejected the *Frye* test, asserting instead that the court should make a "threshold finding of fact with respect to the reliability of the scientific method offered." Without discussing the details of the experts' testimony, the court concluded that the evidence supporting admissibility was credible.[27]

A Delaware trial court held in *State v. Pennell*[28] that DNA evidence was admissible under a state statute similar to the federal rules, but refused to admit probability statistics. There was no dispute about the underlying theory of DNA typing or its general application in the particular case. The defendant challenged the laboratory's claims that the population databank it used was in Hardy-Weinberg equilibrium and that its "binning process" was valid. The defense held that the state's experts' assessment of the probability of declaring a match was overstated. The court accepted some of the defense contentions and faulted the laboratory for its procedure. The state later introduced new evidence based on the laboratory's revised procedure and a new databank. The court agreed to allow the new evidence if the state would provide the raw data to the defendant, but the state did not do so. The court expressed concern over testimony that the measurements of allele size can depend on who is doing the measuring, and it concluded that the state's evidence did not sufficiently support the probability calculation.

Recent Appellate Opinions

As of February 1991, one federal and 10 state appeals from decisions to admit DNA evidence had been decided. Eight of the state appellate courts upheld trial courts' decisions to admit; the other two approved the scientific theory underlying DNA typing, but one excluded the work of a particular laboratory because of process unreliability, and one found that there was sufficient controversy about the methods for assigning statistical weight so that they could not be considered generally accepted. In the sole federal appellate ruling, the Eighth Circuit Court of Appeals reversed a federal trial court's decision to admit DNA typing evidence and directed the lower court to hold a full hearing on admissibility.[29] In the spring and summer of 1990, an intermediate-level appellate court in Texas[30] and the supreme courts of South Carolina,[31] Georgia,[32] North Carolina,[33] and Massachusetts[34] were among the courts that considered the admissibility of DNA evidence. These opinions are of particular interest, because they were issued after sustained debate in the legal and scientific communities about possible flaws in DNA typing technology and possible inadequacies in the population databanks. The courts in Texas, South Carolina, Georgia, and North Carolina upheld the admissibility of DNA evidence; Massachusetts rejected it because of concerns about the adequacy of population genetic interpretation.

In *Kelly v. Texas*, the defendant appealed from a murder conviction, challenging as error the trial court's admission of evidence that compared a semen sample from the crime scene to a blood sample of the suspect. The defendant did not challenge the principles of DNA typing or the general qualifications of the state's five experts. He did attack the methods of the testing laboratory and the statistical expertise of the witnesses. The appellate court was informed that outside experts had twice verified the laboratory's procedures and results. In upholding the trial court's decision to admit the evidence, the appellate court specifically acknowledged the "validity" of the laboratory's techniques.

In July 1990, the Supreme Court of Georgia decided *Caldwell v. State*, a death-penalty case. The appeal grew out of a trial court's decision after a *Frye* hearing (that involved testimony by 10 experts) to admit DNA evidence. Both at the *Frye* hearing and on the appeal, no challenge was made to the scientific principles or general techniques used by the forensic laboratory. The focus was on how the laboratory declared a match between samples, the validity of its probability calculations, and its procedures to ensure quality control. In deciding the appeal, the court first considered whether it was appropriate for the trial court to use a *Frye* hearing to determine whether the laboratory had performed its test with reliable techniques and in an acceptable manner. It concluded that, because of the complexity of the issues and a lack of national standards, the inquiry was appropriate. Although noting that errors, including false positives, could occur, the court ruled that the laboratory's protocol was "adequate to meet these concerns."

The court addressed how the laboratory had conducted a band shift analysis and calculated the power of identity. Despite band shifting, the laboratory had originally decided a match by visual examination. During the course of the trial, as a result of criticism of that technique, it reanalyzed the samples with a monomorphic probe. Such a probe provides an arguably invariant reference point to analyze band shifts across samples. After review of the testimony concerning the reanalysis, the appellate court concluded that this approach to the problem of band shifting was acceptable.

The appellant in *Caldwell* also attacked the calculations that led the testing laboratory to conclude that the chance that a randomly selected person would have the same DNA pattern as that of the sample source and the suspect was 1 in 24,000,000. Only one of the 10 experts had actually examined the laboratory's population databank, and he, a defense witness, insisted that it was not in Hardy-Weinberg equilibrium. The court ruled that, in the absence of supporting testimony, the probability statement generated by the laboratory assumptions could not be accepted. But the court did accept the concept of appropriate statistical calculations, which it erroneously thought did not depend on population theory. (See the discussion of the population genetics question in *Caldwell* in Chapter 3.)

In January 1991, the Supreme Judicial Court of Massachusetts, in *Commonwealth v. Curnin,*[34] became the second state supreme court to refuse to admit DNA typing evidence. After being convicted of rape, in part on the basis of DNA typing evidence, the defendant appealed, arguing that there was no general agreement concerning test methods, use of control samples, or the need for a testing laboratory to meet external performance standards. The high court did not address those arguments, focusing instead on the "lack of inherent rationality" of the process by which the testing laboratory concluded that 1 Caucasian in 59,000,000 would have the DNA pattern represented by the semen stain and the defendant's blood. The court was particularly impressed by the testimony of an expert for the defense who criticized the product rule as unsupported by the laboratory's reference databank, raised the possibility of calculation errors due to ignorance of population substructure, and explained why no assumption would be made as to whether the relevant population was in Hardy-Weinberg equilibrium. Despite its decision to reverse the trial court, the high court made clear that it would not be surprised if the prosecution could correct the weaknesses of its testimony. In the court's words, "it may even be that, by the time of the retrial of this case the prosecution can support the admissibility of evidence of the probability of the alleles disclosed by the DNA test being found elsewhere in the human population. . . ."

Admissibility Statutes

Since 1987, the admissibility of DNA typing evidence was raised repeatedly in the courts, largely in the context of *Frye* hearings. Challenges to admissibility have become more sophisticated over the last 2 years. State legislatures have recently begun to address the matter. Several states have enacted laws that declare that appropriately performed DNA tests are admissible. Although they do not specify what an appropriate test is, the statutes must have been passed with single-locus RFLP analyses by Southern blotting in mind. Arguably, some of them should not be interpreted as applying to technologies that were not in general use and therefore could not have been evaluated by the legislatures that passed the statutes. Such technologies could be validated by amended statutes or by courts in *Frye* or Rule 702 hearings. For most purposes, states with such laws have statutorily resolved disagreements over the scientific reliability of DNA testing, although the questions of whether tests were performed properly in a given case and of the adequacy of statistical calculations based on test results probably remain subject to challenge.

The state laws are of two types. A number of states—including Arkansas, Connecticut, Michigan, Montana, and New Mexico—now specifically admit DNA evidence to assist in the resolution of paternity—noncriminal—cases (and, by inference, probably other disputes concerning biological relationships).[35] Louisiana, Maryland, Minnesota, Virginia, and Washington have enacted laws that recognize the admissibility of DNA evidence in criminal cases.[36]

Maryland requires that the DNA report be delivered to the defendant 2 weeks before the criminal proceeding and specifies that the defendant may require a witness who analyzed the sample to testify as to the chain of custody. The Minnesota statute states that in any civil or criminal trial or hearing DNA evidence is admissible without "antecedent expert testimony that DNA analysis provides a trustworthy and reliable method of identifying characteristics in an individual's genetic material upon a showing that the offered testimony meets the standards for admissibility set forth in the Rules of Evidence"; a companion provision specifically permits the admission of "statistical population frequency evidence . . . to demonstrate the fraction of the population that would have the same combination of genetic markers as was found in a specific human biological specimen." Louisiana provides that "evidence of deoxyribonucleic acid profiles, genetic markers of the blood, and secretor status of the saliva offered to establish the identity of the offender of any crime is relevant as proof in conformity with the Louisiana Code of Evidence."

Legislative interest in DNA evidence remains active, and it is likely that other states will enact laws generally favorable to its admissibility.

DNA DATABANKS ON CONVICTED FELONS: LEGAL ASPECTS

Despite the scientific debate concerning some aspects of DNA typing technology, by late 1990 at least 11 states had implicitly acknowledged its potential value in forensic science by statutorily creating DNA databanks on convicted felons.[37] In general, the laws require that a person convicted of a felony involving a sexual assault submit to phlebotomy before parole; the blood sample is to be subjected to DNA typing and stored under the control of authorities. The California law calls for the testing of felons convicted of murder and other nonsexual felonies involving violence to a person. The Iowa law does not make clear who will be tested. The Virginia law provides for testing of all convicted felons.

Those laws were enacted because of the high rate of repeat felonious behavior by convicted persons. For example, available data on Virginia offenders shows that 36.3% of persons convicted of rape and 32.8% of persons convicted of aggravated assault (including sexual assault) are convicted of another crime within 5 years.[38] The laws are premised on the fact that criminals sometimes leave biological evidence at the crime scene and that the comparison of the results of DNA typing of such samples with profiles stored in the forensic laboratory might lead law-enforcement officials quickly to a prime suspect.

The creation of felon DNA databanks raises a number of challenging constitutional questions, e.g., whether extracting blood for DNA analysis in anticipation of future conduct is an unreasonable search or seizure under the Fourth Amendment and whether the creation of such banks violates a privacy right of the first-degree relatives of persons whose DNA samples are stored (see Chapter 3). This committee is not prepared to recommend how these important questions should be resolved, but recognizes that they deserve careful scrutiny. So far, one federal district court has heard a challenge to the constitutionality of a felon DNA databank. Its order for summary judgment favored the Virginia law.[39]

The committee did not conduct a detailed study of DNA databanks for law-enforcement purposes. However, the committee does recognize that, as scientific and technical concerns about DNA typing are resolved, it is highly likely that databanks will proliferate, interconnect, and communicate. There is clearly a need to conduct further studies on the issue. It will be important to measure the perceived benefits of such databanks against possible harm. We must explore, among other questions, the permissible purposes of such banks, how to minimize invasion of legitimate privacy interests, and how to determine the appropriate response when such interests are violated[40] (see also Chapters 5 and 7).

ASSESSING THE ADMISSIBILITY OF EVIDENCE BASED ON RESULTS OF FURTHER ADVANCES IN DNA TECHNOLOGY

It is important to remember that "DNA typing" is a catch-all phrase for an array of quite different technologies for measuring DNA variations among persons. For some DNA typing methods, the technical basis is well accepted. For others, important scientific questions must be resolved before they are appropriate to use in court.

New developments in DNA technology probably will, and at first should, be the subject of *in limine* hearings (those conducted by a court in deciding on admissibility), as has been the case in recent instances when present technology has been tested. As a general rule, generation of evidence with such new technology should be encouraged if it is adequately supported in court hearings. It is highly desirable that experts in molecular genetics and statistical analysis review new developments and pass on them at a variety of conferences and through published papers. Until there is some consensus in this field, results of using new techniques may not be admitted; a testing period for the new techniques will be needed to determine whether there are unforeseen errors or difficulties, and it will take time to compile the necessary databanks. Otherwise, the normal rules with respect to new developments can be relied on. In fact, new developments should present less difficulty than has been posed by present DNA typing technology, because much of the theory will have already been tested and accepted by the courts.

The issue for courts will be to discern when a technology is so different as to require a full admissibility hearing. Admissibility hearings might be required to evaluate the underlying principle of a scientific method of identification, the particular method for applying the principle, and the performance of a test in a particular case. Regarding the underlying principles, there is, as we have noted, no longer any question concerning the principle that DNA can be used to obtain identification information; admissibility hearings need no longer address the question. Regarding the particular method for applying the principle, the inquiry will depend on the new method or technology. For example, use of a previously unused DNA probe in the context of the basic RFLP technique might require an admissibility hearing on whether the properties of the particular probe (e.g., pattern, sensitivity, or population genetics) are scientifically accepted. Methods of correcting for shifted DNA patterns (that would otherwise fall outside the usual matching rule) might require an admissibility hearing concerning whether the correction procedure has gained scientific acceptance, inasmuch as this substantially changes the method of declaring a match. The use of PCR amplification for sample preparation might require a pretrial hearing on the properties of the technique, because it introduces a novel issue considered by only a few courts thus far—the synthesis of evidence by amplification. And the use of various detection technologies for PCR products might require a pretrial hearing about the characteristics of the detection method and its sensitivity to artifacts. In each case, the court can properly limit inquiry to the substantially novel aspects of the technology, focusing on whether the method is accepted for scientific applications and whether it has been validated for forensic identification. Minor changes in protocols will typically not require pretrial hearings, unless they are likely to affect key issues (such as the matching rule).

SUGGESTIONS FOR USE OF DNA EVIDENCE

Whatever statute or rule of evidence is applicable, some standards for admissibility seem sound to the committee. In view of the importance of DNA typing in both civil and criminal cases, the judge should determine, before allowing DNA evidence to be introduced, that appropriate standards have been followed, that tests were adequately performed by a reliable laboratory, and that the appropriate protocols for DNA typing and formulation of an opinion were fully complied with. In states without relevant statutes, the committee recommends that the court judicially notice the appropriateness of the theoretical basis of DNA typing by using this report, similar reports, and case law. As new methods are used, the courts will have to assure themselves of their validity.

The problem that a court will have to focus on when a standard testing approach is used is not general scientific theory, but actual application. *In limine* hearings can be shortened considerably by stipulations, exchange of data by the parties, and pretrial hearings to avoid unnecessary delay in trials. In the absence of specific objections to laboratory procedures, a court may rely on evidence of accreditation and certifications, a history of adequacy of testing by the laboratory, and other assurances of careful practice. It is not necessary, at this stage of development of DNA typing, to hold extensive admissibility hearings on the general validity of the scientific techniques, although cases will still arise in which the procedures used to report a match will be questioned.

It also might be necessary in a particular case to decide in advance whether an expert will be permitted to characterize the probability of a match in mathematical terms. As noted in Chapter 3, the use of the product rule (which assumes the independence of the frequency distribution of the single-locus probes and is the method by which the likelihood statement is generated) is controversial. At present, courts should take a conservative approach concerning the assumptions underlying the use of the product rule. A considerable degree of discretion and control by the courts in these cases is recommended.

As a general matter, so long as the safeguards we discuss in this report are followed, admissibility of DNA typing should be encouraged. There is no substantial dispute about the underlying scientific principles. However, the adequacy of laboratory procedures and of the competence of the experts who testify should remain open to inquiry. Ultimately, DNA typing evidence should be used without any greater inconvenience than traditional fingerprint evidence.

DNA EVIDENCE AND THE VARIOUS PARTIES IN THE LEGAL SYSTEM

The Jury

Because a jury might overvalue or undervalue scientific evidence, it is appropriate where permitted for the judge to question DNA experts with an eye to aiding the jury. The judge can explain to the jury the role of experts and the role of the jury in evaluating the experts' opinions.

When probability statements are admissible, the judge should not be expected to instruct the jury in detail on how probabilities are computed or how probabilities available from an analysis of DNA material should be combined with probability estimates based on more traditional testimony and other evidence. Those matters are better left to the experts and to the lawyers on summation. The court should encourage the use of charts, written reports, and duplicates of materials that are relied on by the experts, so that the jury can be as well educated as possible in the evaluation of DNA typing evidence. To that end, the court should insist that technical terms be reduced to understandable lay language and that scientific information be presented to the jury in the least confusing form possible.

Special forms of charges are not required. DNA typing may be assessed within the framework of normal forensic laboratory work and can be readily handled with the present rules and forms of charges.

The Prosecutor

The prosecutor will work closely with the investigators and will normally have access to adequately staffed and organized forensic laboratories. The prosecutor should carefully supervise the investigation activities to ensure that DNA typing evidence will be admissible, if it proves relevant.

The prosecutor has a strong responsibility to reveal fully to defense counsel and experts retained by the defendant all material that might be necessary in evaluating the evidence. That includes information on tests that proved inconclusive, on retesting, and on the testing of other persons. Adoption of rules or statutes that require the prosecutor to involve the defense in analysis of DNA samples at the earliest possible moment is highly recommended.

The committee recommends going beyond what is required by the federal rules of criminal procedure and of civil procedure in regard to disclosures concerning DNA evidence. For example, data sheets and other materials obtained from experts who are not designated to testify should be available freely without the need for separate motions, because such materials are important for the evaluation of the scientific evidence in the case of DNA typing. Such free exchange of information, including access to databanks and to samples of evidence DNA, should apply to defense and prosecution experts in both criminal and civil cases.

The Defense

Defense counsel must have access to adequate expert assistance, even when the admissibility of the results of analytical techniques is not in question, because there is still a need to review the quality of the laboratory work and the interpretation of the results. When the prosecutor proposes to use DNA typing evidence or when it has been used in the investigation of the case, an expert should be routinely available to the defendant. If necessary, he or she should be able to apply for funds early in the discovery stages to retain experts without a showing of relevance that might reveal trial strategy. Whenever possible, a portion of the DNA sample should be preserved for independent analysis by the defense.

The prosecutor should promptly reveal to defense counsel that DNA was involved in the investigation and might be available for analysis at the trial. Normally, the criminal-justice system will not provide for the appointment of counsel for the defendant or for payment for experts until the defendant has been arrested or charged. Where a sample of the defendant's tissue is sought for DNA typing, application to the court for DNA experts should be possible even before an arrest has been made.

In our judicial system, jurors are relatively independent. Nevertheless, through limitations on the admissibility of evidence and on the form of its presentation and through the use of a variety of instructions, the court exercises considerable influence. DNA evidence, like other scientific and statistical evidence, can pose special problems of jury comprehension. Courts and attorneys should cooperate to facilitate jury understanding. Innovative techniques, such as allowing jurors to take notes or ask questions, might be considered. Jargon should be avoided, and information should be presented simply, clearly, and fairly. Unless limited by law or court rules, judges should be free to pose questions to witnesses when they feel that the answers might clarify the testimony. Reports and relevant materials should be admitted into evidence so that they can be studied by courts at their leisure. Finally, a judge would not be amiss in pointing out to attorneys the wisdom of including jurors who are found to have a background that enhances their ability to understand the expert testimony.

TESTING LABORATORIES

Other chapters have indicated appropriate standards for the operation of testing laboratories and the collection and analysis of DNA samples. Uniformity in reporting, completeness of reporting (including laboratory protocols and written criteria for interpretation), and stringent quality assurance of laboratories are essential. The court and the jury should have no reason to doubt the accuracy of the processing of information. Laboratories and experts have a particular responsibility to ensure that they are open and candid with the courts. Any reservations about inadequacies or errors should be promptly revealed, and failure to do that should be dealt with seriously. The court should not hesitate to exercise contempt powers and exclude experts who have misled deliberately in the past. Private trade associations and other appropriate groups should also apply pressure to ensure accuracy and candor.

PROTECTIVE ORDERS

Protective orders should not be used to prevent experts on either side from obtaining all relevant information, which can include original materials, data sheets, software protocols, and information about unpublished databanks. A protective order might be appropriate to limit disclosures by attorneys and experts to third parties about proprietary information acquired in the course of a particular case; but as a general rule, any scientific information used in a case should be open to widespread scientific scrutiny. One exception might be when the expert is involved in a current or recently completed study on which he or she does *not* directly rely to develop an opinion. That will ensure that the expert does not lose his or her opportunity to publish as a consequence of testifying. Protective orders to prevent unnecessary intrusion into the privacy of such persons as those who have been cleared after investigation or who are juveniles are appropriate.

AVAILABILITY AND COST OF EXPERTS

Wide use of forensic DNA typing will have considerable costs. Laboratories will be required to be funded by many states and the federal government. The Commonwealth of Virginia, for example, has committed several million dollars to its DNA forensic activities. Costs will be associated with upgrading the databanks when new procedures replace old ones. Increased costs will also be associated with the control, licensing, and oversight of laboratories and technicians. Many experts will need to be available. The defense cost will be substantially increased. Moreover, as DNA typing becomes more generally available, jurors might expect it in situations where it is impossible to produce. A failure to introduce DNA typing evidence could lead to an inference of spoliation, i.e., the destruction or alteration of evidence.

Of course, the early exclusion of suspects who have been cleared by DNA typing evidence will reduce other costs to the judicial system. DNA evidence might also obviate trials in some cases by proving identity fairly conclusively. In general, however, the costs of the criminal-justice system will be increased.

We cannot now accurately estimate the cost of the widespread use of DNA typing, but it can be expected to run into the tens of millions of dollars a year. However, relative to the cost of operating the entire system, the cost of using DNA evidence is minuscule. The quality of justice will be increased by full use of DNA typing. In general, we believe that the expenditures are warranted by the advantages to be expected.

SUMMARY OF RECOMMENDATIONS

Having carefully reviewed the issues, the committee offers the following recommendations:

• Courts should take judicial notice of three scientific underpinnings of DNA typing:

- —The study of DNA polymorphisms can, in principle, provide a reliable method for comparing samples.
- —Each person's DNA is unique (with the exception of identical twins), although the actual discriminatory power of any particular DNA test will depend on the sites of DNA variation examined.
- —The current laboratory procedure for detecting DNA variation (specifically, single-locus probes analyzed on Southern blots without evidence of band shifting) is fundamentally sound, although the validity of any particular implementation of the basic procedure will depend on proper characterization of the reproducibility of the system (e.g., measurement variation) and the inclusion of all necessary scientific controls.

• The adequacy of the method used to acquire and analyze samples in a given case bears on the admissibility of the evidence and should, unless stipulated, be adjudicated case by case. In this adjudication, the accreditation and certification status of the laboratory performing the analysis should be taken into account.

• Because of the potential power of DNA evidence, authorities must make funds available to pay for expert witnesses, and the appropriate parties must be informed of the use of DNA evidence as soon as possible.

• DNA samples (and evidence likely to contain DNA) should be preserved whenever that is possible.

• All data and laboratory records generated by analysis of DNA samples should be made freely available to all parties. Such access is essential for evaluating the analysis.

• Protective orders should be used only to protect the privacy of the persons involved.

REFERENCES AND FOOTNOTES

1. *Frye v. United States*, 293 F.2d 1013, 104 (D.C. Cir. 1923).
2. *Frye v. United States*, 293 F.2d 1013, at 104 (D.C. Cir. 1923).
3. Gianelli PC. The admissibility of novel scientific evidence: *Frye v. United States*, a half-century later. Colum L Rev. 80:1197, 1226, 1980.
4. 144 Misc.2d 956, 545 N.Y.S. 2d 985 (Sup. Ct. 1989).
5. *Id* at 987.
6. 447 N.W. 2d 422 (Minn. 1989).
7. 447 N.W. at 426 (quoting *State v. Carlson*, 267 N.W.2d 170, 176 (Minn. 1980)).
8. *State v. Joon Kyu Kim*, 398 N.W.2d 544 (Minn. 1987).
9. 80 Md. App. 31, 559 A.2d 391 (Md. App. 1989).
10. *United States v. Downing*, 753 F.2d 1224, 1237 (3d Cir. 1985).
11. See, e.g., *Kropinski v. World Plan Executive Council*, 853 F.2d 948, 956 (D.C. Cir. 1988); *Novak v. United States*, 865 F.2d 718 (6th Cir. 1989); *United States v. Smith*, 869 F.2d 348, 352-54 (7th Cir. 1989).
12. *E.g., United States v. Lushen*, 614 F.2d 1164 (8th Cir.), cert. denied, 446 U.S. 939 (1980); *United States v. Williams*, 583 F.2d 1194 (2d Cir. 1978), cert. denied, 439 U.S. 1117 (1979).
13. *United States v. Hendershot*, 614 F.2d 648 (9th Cir. 1980).
14. *United States v. Williams*, 583 F.2d 1194 (2d Cir. 1978), cert. denied, 439 U.S. 1117 (1979).
15. *State v. Hall*, 297 N.W.2d 80 (Iowa 1980).
16. See, e.g., *People v. Marx*, 54 Cal. App.3d 100, 111, 126 Cal. Rep. 350, 356 (1975).
17. See *United States v. Hearst*, 412 F. Supp. 893 (N.D. Cal. 1976), aff'd, 563 F.2d 1331 (1977).
18. See, e.g., *State v. Williams*, 583 F.2d 1194 (2d Cir. 1978), cert. denied, 439 U.S. 1117 (1979) (court's limiting instruction as to spectrographic voice identification stressed that jury could disregard expert testimony and simply listen to the tapes and decide for themselves).
19. 747 F. Supp. 250 (D. Vt. 1990).
20. *Id* at p.263.
21. *Id* at p.263.
22. *United States v. Yee*, ND, Ohio 129 FRD 692 (1990).
23. Superior Court of State of Arizona in the County of Yuma. No. 15589, February 12, 1991.
24. *U.S. v. Porter*, Superior Court of District of Columbia, Criminal Division, FO6277-89 (1991).
25. 533 So.2d 841 (Fla. App. 5 Dist. 1988).
26. 238 Va. 275, 384 S.E.2d 775 (Va. 1989).
27. *Spencer v. Commonwealth*, 238 Va. 295, 384 S.E. 2d 785 (1989).

28. 1989 WL 167430 at 11 (Del. Super. Ct. Nov. 6, 1989) (Gebelein, J.).
29. *U.S. v. Matthew Sylvester Two Bulls*, 918 F.2d 56 (8th Cir. 1990), 925 F. 2d 1127 (8th Cir. 1991) (en banc), vacated after death of defendant. See Weinstein, Rule 702 of the Federal Rules of Evidence is sound; it should be amended (138 F.R.D. 1991) (discussing *Two Bulls*).
30. *Kelly v. Texas*, No. 2089-026-CR (Ct. of Appeals, 2d District, Fort Worth, June 27, 1990).
31. *State v. Ford*, S.C., 392 S.E. 2d. 783 (1990).
32. *Caldwell v. State*, 260 Ga. 278, 393 S.E. 2d. 436 (1990).
33. *State v. Pennington*, 327 N.C. 89, 393 S.E. 2d 847 (1990).
34. *Commonwealth v. Curnin*, 409 Mass. 218, 565 NE 2d 443 (1991).
35. Arkansas Act 723 (1989); Connecticut P.A. No. 89-360 (1989); Michigan Public Act 258 (1989); Montana Code Ann. Sect. 40-5-201 (1989); New Mexico Stat. Ann. Sect. 40-11-5 (1989).
36. Virginia Code Sect. 19.2-270.5; Louisiana Act 340 (1989); Maryland Chap. 430 (1989); Minnesota Stat Sect. 634.23 (1989); Washington Chap. 350 (1989).
37. Cal. Penal Code 290.2; Colo. Code Ann. 17-2-201(g)(I); Ill. Code Ann. 38 1005-4-3; S. Dak. Code Ann. 23-5-14 et seq; Ariz. Code Ann. 31-281; Fla. Code Ann. 943.325; Iowa Code Ann. 13.10; Minn. Code Ann. 609.3461; Nev. Code Ann. 176.111; Wash. Code Ann. 43.43.754; Virg. Code Ann. 19.2-310.2.
38. Report of the Joint Subcommittee Studying Creation of a DNA Test Data Exchange to the Governor and the General Assembly of Virginia. Senate Document No. 29, Commonwealth of Virginia, Richmond, 1990, p.11.
39. *Lawrence R. Jones, et al. v. Edward W. Murray, Director of The Department of Corrections, et al.* W.D. Virginia, Civil No. 90-0572-R. Order for Summary Judgment, March 4, 1991.
40. See, in general, Ballantyne J, Sensabaugh G, Witkowski J., eds. DNA Technology and Forensic Science. Banbury Report 32. Cold Spring Harbor, New York: Cold Spring Harbor Laboratory Press, 1989.

7

DNA Typing and Society

The introduction of any new technology is likely to raise concerns about its impact on society. Financial costs, potential harm to the interests of individuals, and threats to liberty or privacy are only a few of the worries typically voiced when a new technology is on the horizon. DNA typing technology has the potential for uncovering and revealing a great deal of information that most people consider to be intensely private.

The federally established human genome program will yield an unprecedented amount of genetic information and generate new databanks.[1,2] Even apart from the human genome program, DNA technology is moving forward; but this large-scale program, projected to take 10-15 years, is bound to accelerate the acquisition of genetic information. At the same time, it contains a mandate for examining the ethical, social, and legal implications of mapping the human genome, with specific allocation of funds for examining these aspects.[2] A central concern raised by these developments is the safeguarding of the confidentiality of personal genetic information. With greater understanding of the human genome, the potential of misuse of DNA samples collected or preserved for purposes of criminal justice will increase. The more databanks are established, the greater the risk of breaches of confidentiality and misuse of the information.[3]

The social, economic, and ethical concerns in this chapter overlap with the legal aspects addressed in Chapter 6 and the issues in development and use of databanks discussed in Chapter 5.

ECONOMIC ASPECTS

The forensic use of DNA technology will have various economic impacts. The proliferation of DNA evidence in investigations and trials requires a fairly rapid expansion in the number of reliable experts and laboratories. The cost of the equipment, training and proficiency programs, supplies, and personnel will be very large. For example, the three proposed regional laboratories in New York state are estimated to cost $1.4 million per year.[4] The Commonwealth of Virginia has committed several million dollars over the last 3 years to its forensic DNA activities (Paul Ferrara, personal communication, 1990). Material will have to be stored for databanks and for checking suspects. Costs will be associated with the upgrading and changing of databanks when new procedures are adopted. Those costs will affect budgets for police, prosecutors, and courts. Indigent criminal defendants might have a constitutional due-process right to have an expert witness paid for by the government.

The courts themselves must be supplied with reliable assistance in evaluating DNA material. In the federal system, the court can request an expert or panel of experts to assist it, pursuant to Rule 706 of the Federal Rules of Evidence. A special register of scientific experts can be maintained for ready access. The government will generally have to bear this cost. However, if a defendant can afford the cost and asks for expert assistance, the court can assess some costs against the defendant and some against the state.

New costs will also be related to training and certification. The implementation of any new technology requires training and certifying of personnel. Additional costs will be incurred to develop mechanisms to ensure quality control of laboratories that conduct forensic DNA testing.

New technology can grow and make ever larger fiscal demands on society. It is difficult to predict the total cost of DNA testing when it becomes generally available nationwide, but it is reasonable to expect it to amount to tens of millions of dollars a year. That cost is unavoidable, but, given the present fiscal problems at all levels of government, cannot be ignored. Setting up regional and cooperative services is one way of controlling costs. It might not be feasible or appropriate for some small forensic science laboratories to create their own DNA testing capabilities. A major DNA testing center run by the FBI might reduce costs to smaller localities. That potential reduction in monetary cost needs to be balanced against the risks to privacy and confidentiality of having a powerful federal law-enforcement agency in charge of DNA testing and storage of DNA information. If laboratories come to share information, everything could eventually become linked. At the same time, the risks that privacy and confidentiality will be breached might be as great or greater with local control, in that state laws governing the use of criminal records vary widely.

It is likely that the cost of criminal justice will be increased. In some cases, however, early exclusion of suspects who have been cleared by forensic DNA evidence will reduce cost in the judicial system. On balance, the increased costs are small relative to the cost of operating the entire system. The committee believes that the expenditures are warranted by the advantages to be expected.

ETHICAL ASPECTS

Ethical considerations regarding the use of DNA technology in forensic science overlap with various issues addressed in social and legal analyses,[5] including substantive and procedural rights of people and overall nonmonetary costs and benefits likely to result from establishing the use of the new technology in courtroom proceedings.

A threshold question for any ethical inquiry is whether the action or practice under discussion is intrinsically wrong. An action or practice is intrinsically wrong if it violates fundamental ethical principles. These have traditionally been held to include prohibitions against enslavement, torture, gratuitous infliction of harm on human beings, and modes of exploitation that use humans as merely a means (usually without their knowledge or consent) to serve the ends of others.[6] To hold that such actions or practices are intrinsically wrong is not to claim that they can never be justified. For example, if torturing a terrorist who knows the location of a bomb planted to kill a million people is the only way to avert the tragedy, then torture might be justified. That would not yield the conclusion that torture is ethically right, but rather would show that evil acts can sometimes—albeit rarely—be justified as a means of preventing much greater harm.[6]

DNA technology in forensic science is unlikely to violate any fundamental ethical principle of the type described above. Although DNA technology involves new scientific techniques for identifying or excluding people, the techniques are extensions and analogues of techniques long used in forensic science, such as serological and fingerprint examinations, handwriting analyses, photography, and examination of teeth. Ethical questions can be raised about other aspects of this new technology, but it cannot be seen as violating a fundamental ethical principle.

A new practice or technology can be subjected to further ethical analysis by using two leading ethical perspectives. The first examines the action or practice in terms of the rights of people who are affected; the second explores the potential positive and negative consequences (nonmonetary costs and benefits) of the action or practice, in an attempt to determine whether the potential good consequences outweigh the bad.[6]

Moral Rights

Two main questions can be asked about rights: Does the use of DNA technology give rise to any new rights not already recognized? Does the use of DNA technology enhance, endanger, or diminish the rights of anyone who becomes involved in legal proceedings? In answer to the first question, it is hard to think of any new moral rights not already recognized that come into play with the introduction of DNA technology into forensic science. The answer to the second question requires a specification of the classes of people whose rights might be affected and what those rights might be.

The people whose rights might be endangered or diminished seem to be chiefly those who are suspected or accused of or indicted for a crime or involved in other legal proceedings, such as paternity suits, denaturalization, or immigration matters. Does use of DNA technology interfere with or diminish their rights in any way? Might it enhance their rights? Which rights might be endangered?

The current use of DNA technology appears to pose no greater threat to the right to privacy than does normal fingerprinting, placement of photographs in evidence, collection of blood or saliva samples, or other established forensic techniques. DNA technology is not different in principle from those other techniques, although it holds the promise of providing a more definitive identification than most others (fingerprinting is likely to remain the best for a while). If the use of DNA information can be strictly limited to defendant identification, it involves no greater intrusion into the privacy of an accused person than do traditional methods in forensic science, whose aim is to make as definitive an identification as possible. Without strict limits, however, DNA information can be more intrusive into privacy, in that it provides more information about a person.

In some ways, the use of DNA information about suspects can be less intrusive than traditional methods. "Rounding up the usual suspects" by checking a DNA sample against a computerized databank is both much easier and less intrusive than rounding up the suspects themselves. But people who are rounded up are made aware that they are under suspicion and can take protective steps. Where databanks already exist, a fresh blood sample would have to be taken from suspects for confirmation. Thus, it is a complex matter to determine whether the rights of suspects are enhanced or endangered by the use of DNA evidence in the forensic setting, which requires empirical evidence to be subjected to careful analysis.

Concerns about intrusions into privacy and breaches of confidentiality regarding the use of DNA technology in such enterprises as gene mapping are frequently voiced, and they are legitimate ethical worries.[1,6-8] The concerns are pertinent to the role of DNA technology in forensic science, as well as to its widespread use for other purposes and in other social contexts. A potential problem related to the confidentiality of any information obtained is the safeguarding of the information and the prevention of its unauthorized release or dissemination;[5,7] that can also be classified under the heading of abuse and misuse (discussed below), as well as seen as a violation of individual rights in the forensic context.

People have a right not to be wrongly convicted of a crime. To protect that right, a high standard of proof is imposed before a person may be found guilty. In addition, techniques used in gathering and analyzing evidence must have proven reliability (comprising accuracy, precision, specificity, and sensitivity) and should be accepted by a consensus of the scientific community. If DNA technology is as good as or better than other methods used to identify criminals and if the implications and limitations of DNA evidence are recognized by judges and jurors, its use should pose no greater danger to the rights of accused people than the use of currently approved techniques of forensic identification. Moreover, the reliability of DNA evidence will permit it to exonerate some people who would have been wrongfully accused or convicted without it. Therefore, DNA identification is not only a way of securing convictions; it is also a way of excluding suspects who might otherwise be falsely charged with and convicted of serious crimes.

Nonmonetary Costs and Benefits

The ethical perspective by which actions or practices are evaluated in terms of their good and bad consequences is fundamentally sound. Nevertheless, it suffers from both theoretical and practical difficulties.[3] Not only is it difficult to predict good and bad results in advance of gathering sufficient evidence about projected consequences, but it is also sometimes hard to weigh consequences, even if they have already come about. For example, how is it possible to weigh the good consequences of enabling positive identifications to be made with greater certainty by using DNA technology against the bad consequences of drawing mistaken conclusions in particular cases where laboratory techniques or personnel are substandard? Even well-done tests can yield false positives. In approximately 35% of cases performed by the FBI to date, the primary suspect was excluded by DNA (tests on persons who had been prescreened). However, that observation does not resolve the problem of weighing good consequences against bad ones, although it does provide some information that could be used in such weighing.

Another factor to be weighed in a consequentialist ethical analysis is whose interests are to count and whether some people's interests should be given greater weight than others'. For example, there are the interests of the accused, the interests of victims of crime or their families in apprehending and convicting perpetrators, and the interests of society. Whether the interests of society in seeing that justice is done should count as much as the interests of the accused or the victim is open to question. (Here there is an obvious overlap with an ethical analysis from the perspective of rights, and assessment of the consequences of instituting a new practice should include the effects of the new practice on the rights of the people involved.)

Especially when a practice is new and information on projected consequences is scanty, there are problems with relying on balancing the good and bad consequences as a mode of ethical analysis.[9] People who favor one policy or practice predict a balance of good consequences over bad ones, and detractors do the opposite.

One important factor contributing to uncertainty about the use of DNA typing technology is the existence of disagreement among scientific experts.[10] When experts disagree about the use of techniques or statistical methods (such as extrapolations based on population genetics) or about the interpretation of data, the uncertainty is of a different sort from uncertainty that stems simply from scanty evidence drawn from actual consequences. The latter uncertainty can be remedied by gathering more data before a technology is introduced as an accepted standard. If controversy among experts persists, disagreements can erupt whenever empirical evidence is analyzed and specific conclusions are to be drawn.

An overall judgment that DNA technology in forensic science is superior to existing forensic methods requires comparing intersubjectively verified scientific evidence on the reliability and validity of the new method with evidence on the other methods. Certainly, as a personal identification method, fingerprinting is the definitive forensic technique and has many advantages. It has almost 100 years of development, which has established empirically that a person has unique fingerprints; fingerprints can even distinguish between identical twins. Fingerprints are easily detected and developed, and large electronic fingerprint databanks exist all over the world. A fingerprint is a directly observable impression that does not generally involve extensive chemical or biochemical manipulation. Rarely do fingerprint experts differ in conclusions reached after examination of fingerprint evidence. The limitations of the technique are derived from the fact that usable fingerprint evidence is left at crime scenes relatively rarely and can indicate only the presence of a person at a scene.

Another method of identification commonly used in forensic laboratories is forensic serology, i.e., analysis of physiological fluids for genetic markers, such as ABO antigens, enzymes, and serum proteins. The major drawback of these analyses is the degree of specificity provided. The usual battery of serological tests might still allow characterization of a person only as a member of a larger population sharing the same markers. Depending on the panel of tests, the likelihood that a randomly selected person will show the same markers as a person in question can range from 1 in 2 (such as in type O) to 1 in several thousand (such as when many systems are typed and a relatively rare type is found). The literature and case law on paternity disputes suggest that a likelihood of 1 in 20 is reasonably corroborative and that a likelihood of only 1 in 100 can strongly influence the triers of fact.

Thus, although conventional serology can exclude a person, it can also include many members of a population group as the possible origin of a blood, saliva, or seminal fluid stain. Conventional serology is further limited, in that analysis of mixed-fluid stains in which two or more contributors are involved can mask an individual donor. Similarly, only 75-80% of the population are secretors (exhibit their ABO blood type in their other physiological fluids). Thus, the combination of those factors severely limits the power of conventional forensic serological examinations as an individual identifier. Results of serological analysis also are more subjective and can give rise to differing conclusions when interpreted by equally qualified scientists.

Hair evidence is often encountered in sexual assault and other violent crimes. It is valuable as exculpatory evidence and can be informative as to identity, but it lacks specificity. Although hair examiners can associate a hair with racial characteristics and body source (trunk, head, or pubic area) the variations among hairs on a given person make definitive association of a single hair with an individual problematic. The microscopic comparison of hairs is also subjective and can lead to differences of opinion among equally qualified experts. With the advent of DNA technology, especially PCR amplification techniques, the use of hair as an individual identifier will become more common.

Some other forms of individual identification are available to forensic scientists, but have very limited application. For example, examination of teeth is useful in identifying deceased persons or bite marks.

Providing scientific evidence that DNA technology is at least as reliable as other forensic methods and is therefore more likely to result in definitive identification or exclusion of persons suspected or accused of a crime satisfies both the ethical concerns about individual rights and the conditions of an ethical analysis based on weighing good and bad consequences. However, additional assurances are required for the risk-benefit ratio to be favorable in each case in which the technology is used. Therefore, a critical step in accepting the use of DNA technology in criminal trials is establishing safeguards and seeking to prevent abuses.

ABUSE AND MISUSE OF DNA INFORMATION

Even if a technology is scientifically sound and its use is ethically permissible, it is necessary to seek to prevent abuses and misuses in practice. Examples of abuses of DNA technology are unauthorized access to databanks and unauthorized disclosure of information. An example of misuse is the use of DNA information for purposes other than forensic—in other words, going beyond the intended purpose of collecting and storing the information.

A major issue is the preservation of confidentiality of information obtained with DNA technology in the forensic context.[5,7] When databanks are established in such a way that state and federal law-enforcement authorities can gain access to DNA profiles, not only of persons convicted of violent crimes but of others as well, there is a serious potential for abuse of confidential information.[11] The victims of many crimes in urban areas are relatives or neighbors of the perpetrators, and these victims might themselves be former or future perpetrators. There is greater likelihood that DNA information on minority-group members, such as blacks and Hispanics, will be stored or accessed. However, it is important to note that use of the ceiling principle (Chapter 3) removes the necessity to categorize criminals (or defendants in general) by race for the purposes of DNA testing and storage of information in databanks.

Maintaining DNA samples or information about ex-offenders and parolees might be permissible, but requires justification. Even in a felon databank, protections must be instituted. For example, a person's permission should be obtained for the use of his or her DNA information outside the forensic context. If there are no witnesses to a crime, law-enforcement agencies are likely to go directly to the felon databank in their quest for probable suspects. The tendency to use efficient and cost-effective means to solve crimes could result in reducing safeguards, thereby eroding rights of ex-offenders and parolees.

Storage of DNA records of people who have not been convicted of a crime raises ethical questions about the proper "ownership" of such information.[11] DNA information is personal and so should be treated as private, like information in a person's medical record.[8] Outside the forensic context, DNA information should be stored in databanks and released only with the knowledge and explicit permission of the person who is the subject of the information. As for storage of forensic DNA information in databanks, some disagreements remain about propriety and about the prospects for abuse (Chapter 5).

Even when the use of criminal databanks is limited to the local or regional level, the potential for expansion raises questions of misuse. For example, should a whole local population be subject to DNA typing when it is strongly suspected that someone in the population left blood or other fluids at the scene of the crime? Should this be seen as similar to a "frisk" or a simple search that requires a warrant or as an intrusion into someone's body that requires a strong showing of need? The potential for expanded uses of DNA technology that would constitute serious intrusions into the privacy of ordinary citizens requires the setting of guidelines that separate proper use from misuse of the technology.

The release of DNA information on a criminal population for purposes other than law enforcement also constitutes misuse. Employers and insurance companies will certainly have an interest in DNA information on potential employees or customers.[1,8,9] Biomedical and behavioral scientists are likely to want to screen felon databanks and develop new databanks to study various characteristics of convicted offenders. Legal sanctions should be established to deter the unauthorized dissemination or procurement of DNA information that has been obtained for forensic purposes.

EXPECTATIONS

The introduction of a powerful new technology is likely to set up unwarranted or unrealistic expectations. Various expectations regarding DNA typing technology are likely to be raised in the minds of jurors and others in the forensic setting[10] (see Chapter 6).

For example, public perception of the accuracy and efficacy of DNA typing might well put pressure on prosecutors to obtain DNA evidence whenever appropriate samples are available. As the use of the technology becomes widely publicized, juries will come to expect it, just as they now expect fingerprint evidence, surveillance photographs, and audio and visual eavesdropping. Moreover, prosecutors will not want to give defense attorneys the opportunity to ask on summation, "If my client was the perpetrator, where is the DNA evidence?"

Once a prosecutor produces DNA evidence, the defense will be under great pressure to undermine it through the use of reports and experts, because of an assumption that the jury would interpret a failure to call a defense expert as an admission that the DNA evidence is persuasive. Mere cross examination by a defense attorney inexperienced in the science of DNA testing will not be sufficient.

Two aspects of DNA typing technology contribute to the likelihood of its raising inappropriate expectations in the minds of jurors. The first is the jury's perception of an extraordinarily high probability of enabling a definitive identification of a criminal suspect; the second is the scientific complexity of the technology, which results in laypersons' inadequate understanding of its capabilities and failings. Taken together, those two aspects can lead to the jury's ignoring other evidence that it should be considering.

Expectations regarding the power of DNA typing can lead to overlooking or ignoring sources of error or mistakes in applying the technology. For example, jurors' focusing on the probability of correctly identifying a perpetrator might lead them to discount the possibility of laboratory error, whether it stems from incompetence or carelessness of personnel, malfunctioning equipment, or unavoidable mistakes.

The efficacy and accuracy of a new technology typically are initially demonstrated by the most highly competent and knowledgeable practitioners. As DNA typing becomes routine, the quality of laboratories and personnel using it might decrease while still meeting the standards required for accreditation or licensing. However, the expectations of judges and juries might remain high, because of the superior knowledge and competence of the initiators of the technology. Later gains in experience and improved typing could lead to an increase in quality.

As large felon databanks are created, the forensic community could well place more reliance on DNA evidence, and a possible consequence is the underplaying of other forensic evidence. Unwarranted expectations about the power of DNA technology might result in the exclusion of relevant evidence.

Both prosecutors and defense counsel are entitled to benefit from the power of DNA evidence, but they should not oversell it. DNA evidence is not infallible; all laboratory work is subject to error; and, given current population databanks and laboratory protocols, a witness or prosecutor will seldom (if ever) be justified in stating that the probability that a reported DNA match involves someone other than the suspect is so low as to make that possibility entirely implausible. Claims that treat DNA identifications as though they are as reliable as fingerprint identifications in the typical rape or murder case are unjustified; until technology and databanks improve, they are likely to remain so.

Presentations suggesting to a judge or jury that DNA typing is infallible can rarely be justified and should generally be avoided. However, there might be instances where a prosecutor could legitimately argue that the DNA evidence conclusively proves that the defendant committed the offense. Two examples are illustrative:

- The victim is confined to an institution where access is limited to relatively few male attendants. Semen taken from the vagina is subjected to analysis and compared to blood samples from all possible males with access to the victim. The defendant's known sample is the only profile that matches the evidentiary sample. In this circumstance, the prosecutor could well argue that only the defendant could have committed the crime.
- In a prosecution for sexual assault of a child, again a limited number of people might have access to the child, with only one possible donor matching the evidentiary sample. Again, the prosecutor might argue that the DNA evidence is conclusive.

ACCOUNTABILITY AND PUBLIC SCRUTINY

Because the application of DNA typing in forensic science is to be used in the service of justice, it is especially important for society to establish mechanisms for accountability and to ensure appropriate public scrutiny.

Accountability must be an issue in proficiency testing and accreditation. There is reason to be skeptical of entrusting any important regulatory matters to a self-regulating organization. Accordingly, any organization conducting accreditation or regulation of DNA technology for forensic purposes should be free of influence of private companies, public laboratories, or other organizations actually engaged in laboratory work.

Private laboratories used for testing should not be permitted to withhold information from defendants on the grounds that "trade secrets" are involved. Alternatively, law-enforcement agencies could use only public laboratories for testing, so that the issue of "trade secrets" would not arise.[10] Critics of DNA testing have suggested that the profit motive of private testing companies undermines their reliability. Although that criticism might be justified when companies are eager to market a product before it is ready, no general indictment of private companies on this basis is justified.

Testing methods and data need to be made available for public scrutiny. There has been a notable dearth of published research in forensic DNA testing by scientists unconnected to the companies that market the tests. In contrast with the research approach whereby new drugs and biomedical devices undergo controlled trials of safety and efficacy, forensic science has used more informal modes of evaluating new techniques. The process of peer review used to assess advances in biomedical science and technology should be used for forensic DNA technology.

Whether in publications or in court, companies might be reluctant to reveal their specific testing methods or the population data used to determine the probability of a match, because they consider this information to constitute a trade secret that could be exploited by competitors. However, the integrity of the scientific method and judicial due process demand that such information be revealed, particularly in criminal cases. The scientific community should require that the same standards used to assess new findings in other sectors of science be applied to DNA typing in the forensic setting.

INTERNATIONAL EXCHANGE

The need for international cooperation in law enforcement calls for appropriate scientific and technical exchange among nations. As in other areas of science and technology, dissemination of information about DNA typing and training programs for personnel likely to use the technology should be encouraged. It is desirable that all nations that will collaborate in law-enforcement activities have similar standards and practices, so efforts should be furthered to exchange scientific knowledge and expertise regarding DNA technology in forensic science.

SUMMARY OF RECOMMENDATIONS

• In the forensic context as in the medical setting, DNA information is personal, and a person's privacy and need for confidentiality should be respected. The release of DNA information on a criminal population without the subjects' permission for purposes other than law enforcement should be considered a misuse of the information, and legal sanctions should be established to deter the unauthorized dissemination or procurement of DNA information that was obtained for forensic purposes.

• Prosecutors and defense counsel should not oversell DNA evidence. Presentations that suggest to a judge or jury that DNA typing is infallible are rarely justified and should be avoided.

• Mechanisms should be established to ensure the accountability of laboratories and personnel involved in DNA typing and to make appropriate public scrutiny possible.

• Organizations that conduct accreditation or regulation of DNA technology for forensic purposes should not be subject to the influence of private companies, public laboratories, or other organizations actually engaged in laboratory work.

• Private laboratories used for testing should not be permitted to withhold information from defendants on the grounds that trade secrets are involved.

• The same standards and peer-review processes used to evaluate advances in biomedical science and technology should be used to evaluate forensic DNA methods and techniques.

• Efforts at international cooperation should be furthered to ensure uniform international standards and the fullest possible exchange of scientific knowledge and technical expertise.

REFERENCES

1. U.S. Congress, Office of Technology Assessment. Mapping our genes—the genome projects: how big, how fast? Washington, D.C.: U.S. Government Printing Office, 1988.
2. U.S. Department of Health and Human Services and U.S. Department of Energy. Understanding our genetic inheritance: the U.S. Human Genome Project, the first five years FY 1991-1995. Springfield, Virginia: National Technical Information Service, 1990.
3. President's Commission for the Study of Ethical Problems in Medicine and Biomedical

and Behavioral Research. Screening and counseling for genetic conditions. Washington, D.C.: U.S. Government Printing Office, 1983.

4. DNA Report of New York State Forensic Analysis Panel. Albany, New York, 1989.
5. U.S. Congress, Office of Technology Assessment. Genetic witness: forensic uses of DNA tests. Chapters 3-5. OTA-BA-438. Washington, D.C.: U.S. Government Printing Office, 1990.
6. Beauchamp TL, Childress JF. Principles of biomedical ethics. Chapter 2. 3rd ed. New York: Oxford University Press, 1989.
7. de Gorgey A. The advent of DNA databanks: implications for information privacy. Am J Law Med. 16:381-398, 1990.
8. Macklin R. Mapping the human genome: problems of privacy and free choice. Pp. 107-114 in: Milunsky A, Annas GJ, eds. Genetics and the law. III. New York: Plenum Press, 1984.
9. President's Commission for the Study of Ethical Problems in Medicine and Biomedical and Behavioral Research. Splicing life. Washington, D.C.: U.S. Government Printing Office, 1982.
10. Annas GJ. DNA fingerprinting in the twilight zone. Hastings Center Rep. 20:35-37, March/April 1990.
11. National Association of Attorneys General. Resolution, adopted at winter meeting. December 10-13, 1989, Phoenix, Arizona.

Organizational Abbreviations

AFIS	Automated Fingerprint Identification Systems
ASCLD	American Society of Crime Laboratory Directors
ASCLD-LAB	American Society of Crime Laboratory Directors—Laboratory Accreditation Board
ASHG	American Society of Human Genetics
CACLD	California Association of Crime Laboratory Directors
CAP	College of American Pathologists
DHHS	U.S. Department of Health and Human Services
DOJ	U.S. Department of Justice
FBI	Federal Bureau of Investigation
NCIC	National Crime Information Center
NIH	National Institutes of Health
NIJ	National Institute of Justice
NIST	National Institute of Standards and Technology
TWGDAM	Technical Working Group on DNA Analysis Methods

Glossary

A single-letter designation of the purine base adenine; also used in diagrams to represent a nucleotide containing adenine

Adenine a purine base; one of the four nitrogen-containing molecules present in nucleic acids DNA and RNA; designated by the letter A

Allele one of two or more alternative forms of a gene

Allele frequency the proportion of a particular allele among the chromosomes carried by individuals in a population

AMP-FLP amplified fragment length polymorphism

Autoradiogram (autoradiograph; autorad) a photographic recording of the positions on a film where radioactive decay of isotopes has occurred

Autosome any of the chromosomes other than the sex chromosomes, X and Y

Band the visual image representing a particular DNA fragment on an autoradiogram

Band shift the phenomenon in which DNA fragments in one lane of a gel migrate at a rate different from that of identical fragments in other lanes of the same gel

Basepair two complementary nucleotides held together by hydrogen bonds; basepairing occurs between A and T and between G and C

Biallelic (see *Diallelic*)

Blot see Southern blot

C single-letter designation of the pyrimidine base cytosine; also used in diagrams to represent a nucleotide containing cytosine

Chromosome the structure by which hereditary information is physically transmitted from one generation to the next; the organelle that carries the genes

Controls tests performed in parallel with experimental samples and designed to demonstrate that a procedure worked correctly

Cytosine a pyrimidine base; one of the four nitrogen-containing molecules in nucleic acids DNA and RNA; designated by the letter C

Degradation the breaking down of DNA by chemical or physical means

Denaturation the process of unfolding of the complementary double strands of DNA to form single strands

Deoxyribonucleic acid (DNA) the genetic material of organisms, usually double-stranded—composed of two complementary chains of nucleotides in the form of a double helix; a class of nucleic acids characterized by the presence of the sugar deoxyribose and the four bases adenine, cytosine, guanine, and thymine

Diallelic DNA variation showing only two forms with a frequency of more than 1%

Diploid having two sets of chromosomes, in pairs (compare haploid)

DNA deoxyribonucleic acid

DNA band the visual image representing a particular DNA fragment on an autoradiogram

DNA databank (database) a collection of DNA typing profiles of selected or randomly chosen individuals

DNA polymerase an enzyme that catalyzes the synthesis of double-stranded DNA

DNA probe a short segment of single-stranded DNA labeled with a radioactive or chemical tag that is used to detect the presence of a particular DNA sequence through hybridization to its complementary sequence

Electrophoresis a technique in which different molecules are separated by their rate of movement in an electric field

Enzyme a protein that is capable of speeding up a specific chemical reaction but which itself is not changed or consumed in the process; a biological catalyst

Ethidium bromide an organic molecule that binds to DNA and fluoresces under ultraviolet light and is used to identify DNA

G single-letter designation of the purine base guanine; also used in diagrams to represent a nucleotide containing guanine

Gamete a haploid reproductive cell

Gametic (phase) equilibrium the state at loci on different chromosomes when the allele at one locus in the gamete varies independently of that at the other loci; in gametic (phase) disequilibrium, a specific allele at one locus is associated with an allele at another locus on a different chromosome with a frequency greater than expected by chance (see linkage disequilibrium)

Gel semisolid matrix (usually agarose or acrylamide) used in electrophoresis to separate molecules

Gene the basic unit of heredity; a sequence of DNA nucleotides on a chromosome

Gene frequency the relative occurrence of a particular allele in a population

Genetic drift random fluctuation in allele frequencies

Genome the total genetic makeup of an organism

Genotype the genetic makeup of an organism, as distinguished from its physical appearance or phenotype

Guanine a purine base; one of the four nitrogen-containing molecules present in nucleic acids DNA and RNA; designated by the letter G

Haploid having one set of chromosomes (compare diploid)

Hardy-Weinberg equilibrium the condition, for a particular genetic locus and a particular population, with the following properties: allele frequencies at the locus are constant in the population over time and there is no statistical correlation between the two alleles possessed by individuals in the population; such a condition is approached in large randomly mating populations in the absence of selection, migration, and mutation

Heredity the transmission of characteristics from parent to offspring

Heterozygote a diploid organism that carries different alleles at one or more genetic loci on its homologous chromosomes

Heterozygous having different alleles at a particular locus; for most forensic DNA probes, the autoradiogram displays two bands if the person is heterozygous at the locus

HLA see human leukocyte antigen

Homology similarity between two structures or functions indicative of a common evolutionary origin

Homozygote a diploid organism that carries identical alleles at one or more genetic loci on its homologous chromosomes

Homozygous having the same allele at a particular locus; for most forensic DNA probes, the autoradiogram displays a single band if the person is homozygous at the locus

Human leukocyte antigen (HLA) protein-sugar structures on the surface of most cells, except blood cells, that differ among individuals and are important for acceptance or rejection of tissue grafts or organ transplants; the locus of one particular class, HLA DQα, is used for forensic analysis with PCR

Hybridization the reassociation of complementary strands of nucleic acids, nucleotides, or probes

Isotope an alternative form of a chemical element; used particularly in reference to the radioactive alternative forms, or radioisotopes

Linkage disequilibrium the phenomenon in which a specific allele at one locus is non-randomly associated with an allele at another locus

Locus (pl. loci) the specific physical location of a gene on a chromosome

Marker a gene with a known location on a chromosome and a clear-cut phenotype that is used as a point of reference in the mapping of other loci

Membrane the matrix (usually nylon) to which DNA is transferred during the Southern blotting procedure

Molecular-weight size marker DNA fragments of known size, from which the size of an unknown DNA sample can be determined

Monomorphic probe a probe that detects the same allele and hence the same pattern in everyone

Multilocus probe a DNA probe that detects genetic variation at multiple sites; an autoradiogram of a multilocus probe yields a complex, stripe-like pattern of 30 or more bands per individual

Mutagen a physical agent (e.g., x rays) or chemical agent that induces changes in DNA

Nucleic acid a nucleotide polymer of which major types are DNA and RNA

Nucleotide a unit of nucleic acid composed of phosphate, a five-carbon sugar (ribose or deoxyribose), and a purine or a pyrimidine base

PCR polymerase chain reaction

Phenotype the physical appearance or functional expression of a trait

Point mutation an alteration of one nucleotide in chromosomal DNA that consists of addition, deletion, or substitution of nucleotides

Polymerase chain reaction (PCR) an in vitro process that yields millions of copies of desired DNA through repeated cycling of a reaction that involves the enzyme DNA polymerase

Polymorphism the presence of more than one allele of a gene in a population at a frequency greater than that of a newly arising mutation; operationally, a population in which the most common allele at a locus has a frequency of less than 99%

Population a group of individuals occupying a given area at a given time

Probe a short segment of single-stranded DNA tagged with a reporter molecule, such as radioactive phosphorus atom, that is used to detect a particular complementary DNA sequence

Proficiency tests tests to evaluate the competence of technicians and the quality performance of a laboratory; in open tests, the technicians are aware that they are being tested, but in blind tests, they are not aware; internal proficiency tests are conducted by the laboratory itself, and external tests are conducted by an agency independent of the laboratory being tested

Protein a chain of amino acids joined by peptide bonds

Purine the larger of two kinds of bases found in DNA and RNA; a nitrogenous base with a double-ring structure, such as adenine or guanine (compare pyrimidine)

Pyrimidine the smaller of two kinds of bases found in DNA and RNA; a nitrogenous base with a single-ring structure, such as cytosine, thymine, and uracil (compare purine)

Quality assurance a program conducted by a laboratory to ensure accuracy and reliability of tests performed

Quality control internal activities or activities according to externally established standards used to monitor the quality of DNA typing to meet and satisfy specified criteria

Recombinant DNA fragments of DNA from two different species, such as a bacterium and a mammal, spliced into a single molecule

Replication the synthesis of new DNA from existing DNA

Restriction endonuclease, restriction enzyme an enzyme that cleaves DNA molecules at particular base sequences

Restriction fragment length polymorphism (RFLP) variation in the length of DNA fragments produced by a restriction endonuclease that cuts at a polymorphic locus

RFLP restriction fragment length polymorphism

RFLP analysis technique that uses single-locus or multi-locus probes to detect variation in a DNA sequence according to differences in the length of fragments created by cutting DNA with a restriction enzyme

Ribonucleic acid (RNA) a class of nucleic acids characterized by the presence of the sugar ribose and the pyrimidine uracil, as opposed to the thymine of DNA

RNA ribonucleic acid

Serology the discipline concerned with the immunologic study of body fluids

Serum the liquid that separates from blood after coagulation

Sex chromosomes (x and y chromosomes) chromosomes that are different in the two sexes and that are involved in sex determination

Sex-linked characteristic a genetic characteristic, such as color blindness, that is determined by a gene on a sex chromosome and shows a different pattern of inheritance in males and females; X-linked is a more specific term

Single-locus probe a DNA probe that detects genetic variation at only one site in the genome; an autoradiogram that uses one single-locus probe usually displays one band in homozygotes and two bands in heterozygotes

Somatic cells the differentiated cells that make up the body tissues of multicellular plants and animals

Southern blot the nylon membrane to which DNA adheres after the process of Southern blotting

Southern blotting the technique for transferring DNA fragments that have been separated by electrophoresis from the gel to a nylon membrane

Standards criteria established for quality control and quality assurance; established or known test reagents, such as molecular-weight standards

T single-letter designation of the pyrimidine base thymine; also used in diagrams to represent a nucleotide containing thymine

Tandem repeats multiple copies of an identical DNA sequence arranged in direct succession in a particular region of a chromosome

Taq polymerase a DNA polymerase used to form double-stranded DNA from nucleotides and a single-stranded DNA template in the PCR technique

Thymine a pyrimidine base; one of the four nitrogen-containing molecules present in nucleic acids DNA and RNA; designated by the letter T

Uracil a pyrimidine in RNA

Variable number of tandem repeats (VNTR) repeating units of a DNA sequence for which the number varies between individuals

VNTR variable number of tandem repeats

Zygote diploid cell that results from the fusion of male and female gametes

Biographical Information on Committee Members

VICTOR A. McKUSICK (Chairman) is University Professor of medical genetics at Johns Hopkins. He received his M.D. from the Johns Hopkins University School of Medicine. Dr. McKusick is the founding coeditor of the international journal *GENOMICS* and served as founding president of the Human Genome Organization (HUGO). Dr. McKusick is a member of the National Academy of Sciences and served as a member of the Academy's Committee on Mapping and Sequencing the Human Genome. He also belongs to the American Society for Clinical Investigation, Association of American Physicians, American Society of Human Genetics, American Philosophical Society, American Academy of Arts and Sciences, Royal College of Physicians (London), and Académie Nationale de Médecine (France).

C. THOMAS CASKEY is professor of medicine, biochemistry, and cell biology, at Baylor College of Medicine and Henry and Emma Meyer Professor in molecular genetics, director of the Institute for Molecular Genetics, and investigator of the Howard Hughes Medical Institute. He received his M.D. from Duke University. His research focuses on inherited disease and mammalian genetics. Dr. Caskey's professional affiliations include membership in the Institute of Medicine, National Academy of Sciences, American Society of Human Genetics, Human Genome Organization, and Association of American Physicians. He served as chairman of the Advisory Panel on Forensic Uses of DNA Tests, U.S. Congress Office of Technology Assessment. Dr. Caskey resigned from the committee on December 21, 1991.

PAUL B. FERRARA is director of the Virginia Division of Forensic Science. He holds Ph.D. degrees in organic chemistry from Syracuse University and from the State University of New York. After working as a research chemist for du Pont, Dr. Ferrara entered the field of forensic science in the Northern Virginia Police Laboratory, which became the Northern Virginia Regional Facility of the statewide Forensic Laboratory System. He is a charter member and past president of the Mid-Atlantic Association of Forensic Scientists, a member of the Executive Board of the American Society of Crime Laboratory Directors (ASCLD), and chairman of the DNA Implementation Committee of ASCLD.

MICHAEL W. HUNKAPILLER is vice president for science and technology and general manager of Applied Biosystems Inc. He holds a Ph.D. in chemical engineering from the California Institute of Technology. His research focuses on the automation of procedures used in the structural analysis and synthesis of proteins and DNA. Dr. Hunkapiller is the author of more than 100 publications and the inventor on more than 20 patents. His current professional activities include service on the editorial boards of *Technique*, *Genomics*, and *Analytical Biochemistry*, and he is a member of the American Association for the Advancement of Science, American Chemical Society, and Human Genome Organization. Dr. Hunkapiller resigned from the committee on August 17, 1990.

HAIG H. KAZAZIAN, JR. serves as director of the Center of Medical Genetics at Johns Hopkins University School of Medicine. He is the Frank Sutland Professor of pediatric genetics and professor of medicine and gynecology and obstetrics at the medical school and professor of biology at the university. His research has concentrated on mutation analysis in genetic diseases and he has had a long interest in DNA diagnosis of genetic disease. Dr. Kazazian has served on numerous NIH committees and editorial boards. He is a member of the American Pediatric Society, American Society of Clinical Investigation, Association of American Physicians, and the American Society of Human Genetics.

MARY-CLAIRE KING is professor of epidemiology in the School of Public Health of the University of California, Berkeley and of genetics in the Department of Molecular and Cell Biology. She holds a Ph.D. in genetics from the University of California, Berkeley. Dr. King's expertise is in human genetics, genetic epidemiology, and population genetics. Her professional affiliations include membership in many NIH committees and the American Association for the Advancement of Science, American Epidemiological Society, Society for Epidemiological Research, and American Society of Human Genetics.

ERIC S. LANDER is a member of the Whitehead Institute for Biomedical Research, associate professor of biology at the Massachusetts Institute of Technology, and director of the MIT Center for Genome Research. He received his A.B. in mathematics from Princeton University and his Ph.D. in mathematics from Oxford University. Dr. Lander's expertise is in human and mouse genetics, molecular biology, population genetics, mathematics, and statistics. He has been a MacArthur Fellow for research in human genetics and a Rhodes Scholar. He has been a member of numerous NIH committees and has chaired the Genome Research Review Committee for the National Center for Human Genome Research. He also served on the Advisory Panel on Forensic Uses of DNA Tests for the U.S. Congress Office of Technology Assessment. Other professional activities include service on the editorial boards of *GENOMICS*, *Mammalian Genome*, *Theoretical Population Biology*, *BioTechniques*, and *PCR Methods and Applications*.

HENRY C. LEE serves as director of the Forensic Science Laboratory of the Connecticut State Police and professor of forensic science at the University of New Haven. He received an M.S. and a Ph.D. in biochemistry from New York University. Dr. Lee has been editor of the *Hua-Lian Daily News* and captain at the Taipei Police Headquarters in Taiwan. He has conducted special training activities in the various components of forensic science. His professional memberships include those in the American Academy of Forensic Science, American Society of Crime Laboratory Directors, International Association of Forensic Science, Forensic Science Society (England), New York Academy of Science, and International Homicide Detective Association.

RICHARD O. LEMPERT is Francis A. Allen Collegiate Professor of Law and professor of sociology at the University of Michigan. Dr. Lempert has served on National Science Foundation panels on law and social science and on the global effects of environmental change. He has also served as a member, vice chair, and chair of the National Research Council's Committee on Law and Justice. He is a member of the American Sociological Association and the Law and Society Association and has served the latter as editor of *Law and Society Review*. He is now a trustee of the Law and Society Association and a member of the editorial boards of *Violence and Victims* and *The Law and Society Review*.

RUTH MACKLIN serves as professor of bioethics in the Department of Epidemiology and Social Medicine at the Albert Einstein College of Medicine. She received her M.A. and Ph.D. in philosophy from Case Western Reserve University. Dr. Macklin has published on medical ethics, and her list of honors and awards includes an American Council of Learned Societies fellowship and elected membership in the Institute of Medicine.

THOMAS G. MARR serves as senior staff investigator at the Cold Spring Harbor Laboratory. He received his M.S. and Ph.D. in biology from Michigan State University. The subjects of Dr. Marr's research studies include molecular genetics analysis and databases. He is a member of the Human Genome Joint Informatics Task Force, NIH/DOE, as well as the Technical Advisory Group-Genome Database, Howard Hughes Medical Institute/The Johns Hopkins University. Dr. Marr sits on the editorial board of *Mammalian Genome*, and his other professional affiliations include membership in the American Statistical Association, Association for Computing Machinery, and Institute of Electrical and Electronic Engineers.

PHILIP R. REILLY is executive director of the Shriver Center for Mental Retardation. He received his J.D. from Columbia University and his M.D. from Yale University. Dr. Reilly has written numerous articles and several books, including *Genetics, Law, and Social Policy.* He lectures on ethical issues in genetic testing, paternity litigation, and DNA forensic science. Dr. Reilly also served as a member of the National Research Council Committee on Public Information in the Prevention of Occupational Cancer.

GEORGE F. SENSABAUGH, JR., is professor of forensic science and biomedical sciences at the University of California, Berkeley. He received his B.A. in philosophy from Princeton University and a doctorate in criminalistics from the University of California, Berkeley. Dr. Sensabaugh serves on the editorial boards of the *Journal of Forensic Science*, *American Journal of Forensic Medicine and Pathology*, *Forensic Science Reviews*, and *Journal of Forensic Science Society (Great Britain)*. He is a member of the American Association for the Advancement of Science, American Academy of Forensic Science, International Society for Forensic Hemogenetics, American Society of Human Genetics, New York Academy of Science, and American Chemical Society.

JACK B. WEINSTEIN is U.S. district judge for the Eastern District of New York. He has published treatises, books, and articles, primarily on evidence and procedures. He is adjunct professor at the Columbia and Brooklyn law schools. He served on the National Research Council Committee on Statistics and Law.

Participants

NORMAN ARNHEIM, University of Southern California
MICHAEL BAIRD, Lifecodes Corporation
BRUCE BUDOWLE, Federal Bureau of Investigation
ROBERT M. COOK-DEEGAN, Georgetown University
ROBIN COTTON, Cellmark Diagnostics
HENRY ERLICH, Cetus Corporation
SIMON FORD
DANIEL L. HARTL, Washington University School of Medicine
ROGER KAHN, Miami Metro Dade Police Department
Y. W. KAN, University of California, San Francisco
KENNETH KIDD, Yale University
ALAN LEVY, Fort Worth Tarrant County District Attorney's Office
GARY MARX, Massachusetts Institute of Technology
LAURENCE D. MUELLER, University of California, Irvine
THOMAS H. MURRAY, Case Western Reserve University
PETER J. NEUFELD
DENNIS REEDER, National Institute of Standards and Technology
BARRY SCHECK, Benjamin N. Cardozo School of Law
CHARLES TAYLOR, University of California, Los Angeles
BRUCE WEIR, North Carolina State University

Index

G

H

I

J

K

L